Paramilitarism and Neoliberalism

Paramilitarism and Neoliberalism

Violent Systems of Capital Accumulation in Colombia and Beyond

Jasmin Hristov

PLUTO PRESS

First published 2014 by Pluto Press
345 Archway Road, London N6 5AA

www.plutobooks.com

British Library Cataloguing in Publication Data
A catalogue record for this book is available from the British Library

ISBN   978 0 7453 3501 8   Hardback
ISBN   978 0 7453 3700 5   Paperback
ISBN   978 1 7837 1232 8   PDF eBook
ISBN   978 1 7868 0106 7   Kindle eBook
ISBN   978 1 7868 0105 0   EPUB eBook

Library of Congress Cataloging in Publication Data applied for

10   9   8   7   6   5   4   3   2   1

Typeset by Stanford DTP Services, Northampton, England
Text design by Melanie Patrick

I dedicate this work to the memory of all those who have died at the hands of the paramilitary and members of the Colombian armed forces, police and special units.

Ni perdon, ni olvido

Contents

List of Tables

List of Figures

List of Acronyms

ACCU	Autodefensas Campesinas de Córdoba y Urabá (Peasant Self-defense of Córdoba and Urabá)
ACVC	Asociación Campesina del Valle del Río Cimiterra
ADN	Alianza Democratica Nacional (Democratic National Alliance)
AI	Amnesty International
ANDI	Asociación Nacional de Empresarios de Colombia (National Association of Industrialists)
ANIC	Agencia Nacional de Inteligencia Colombiana (National Colombian Agency of Intelligence)
ANTHOC	Asociación Nacional de Trabajadores Hospitalarios y deClínicas (National Association of Hospital and Clinic Workers)
ANUC	Asociación Nacional de Usuarios Campesinos (National Association of Peasants)
ANZORC	Asociación Nacional de Zonas de Reservas Campesinas
ASCAMCAT	Asociación Campesina del Catatumbo
ASOPEMA	Asociación de Pequeños y Medianos Agricultores del Tolima (Association of Small and Medium-Scale Farmers of Tolima)
ASTDEMP	Asociación Santandereana de Servidores Publicos (Santander Association of Public Servants)
ASTRACATOL	Asociación de Trabajadores Campesinos de Tolima (Association of Peasant Workers of Tolima)
ASTRACAVA	Asociación de Trabajadores Campesinos del Valle de Cauca
AUC	Autodefensas Unidas de Colombia (United Self-defense of Colombia)
BACRIM	bandas criminals (criminal gangs)
CARSI	Central American Regional Security Initiative
CCJ	Comisión Colombiana de Juristas
CELAC	Comunidad de Estados Latinoamericanos y Caribeños

CINEP	Centro de Investigación y Educación Popular (Center for Popular Research and Education in Colombia)
CISCA	Comite Intersectorial del Catatumbo
CJL	Corporación Jurídica Libertad (Juridic Corporation Liberty)
CODHES	Consultoria para los Derechos Humanos y el Desplazamiento (Consultancy on Human Rights and Displacement)
CONAP	Coordinación Nacional de Organizaciones Agrarias y Populares
CONVIVIR	Servicios Especiales de Vigilancia y Seguridad Privada
CRIC	Consejo Regional Indígena del Cauca (Indigenous Regional Council of Cauca)
CTI	Cuerpo Tecnico de Investigación (Technical Investigative Unit)
CUT	Central Unitaria de Trabajadores de Colombia (Confederation of Colombian Workers)
DANE	Departamento Nacional Administrativo de Estadisticas (National Administrative Department of Statistics)
DAS	Departamento Administrativo de Seguridad (Administrative Department of Security)
ECOPETROL	Empresa Colombiana de Petróleos S.A. (Colombian Petroleum Company)
ELN	Ejército de Liberación Nacional (National Liberation Army)
ENS	Escuela Nacional Sindical (National Trade Union School)
ESMAD	Escuadron Movil Antidisturbios (Anti-Disturbances Mobile Squadron)
EZLN	Ejército Zapatista de Liberacion Nacional (Zapatista Army of National Liberation)
FALANGE	Fuerzas Armadas de Liberacion Anti-comunista de Guerras de Eliminacion (Anti-communist Liberation Armed Forces of Wars of Elimination)
FAO	Food and Agriculture Organization (UN)
FARC	Fuerzas Armadas Revolucionarias de Colombia (Revolutionary Armed Forces of Colombia)

FARC-EP	Fuerzas Armadas Revolucionarias de Colombia— Ejército del Pueblo (Revolutionary Armed Forces of Colombia—People's Army)
FDI	foreign direct investment
FECODE	Federacion Colombiana de Educadores (Colombian Federation of Educators)
FEDERACAFE	Asociación Nacional de Cafeteros (National Federation of Coffee Growers)
FENALCO	Federacion Nacional de Comerciantes (National Federation of Traders)
FENSUAGRO	Federacion National Sindical Unitaria Agropecuaria (Federation of Agricultural Workers and Small-Scale Farmers)
FINAGRO	Fondo para el Financiamento del Sector Agropecuario
FTA	free trade agreement
GAULA	Grupos de Acción Unificada por la Libertad Personal (Joint Action Units for Personal Liberty)
HDI	Human Development Index
HRW	Human Rights Watch
IDB	Inter-American Development Bank
ICFTU	International Confederation of Free Trade Unions
ICTUR	International Centre for Trade Union Rights
IIE	Institute for International Economics
IMF	International Monetary Fund
INCODER	Instituto Colombiano de Desarollo Rural (Colombian Institute for Rural Development)
INCORA	Instituto Colombiano para la Reforma Agraria (Colombian Institute for Agrarian Reform)
ISA	International Sociological Association
ISI	import substitution industrialization
JUCO	Juventud Comunista Colombiana
MAS	Muerte a Secuestradores (Death to Kidnappers)
MIA	Mesa de Interlocución Agraria Nacional
MLN	Movimiento de Liberacion Nacional (Movement of National Liberation)
MNCs	multinational corporations
MOVICE	Movimiento Nacional de Víctimas de Crimenes de Estado (National Movement of Victims of State-sponsored Crimes)

NODAL	Noticias de Latinoamerica y el Caribe
OAS	Organization of American States
OCENSA	Oleoducto Central S.A.
OFP	Organizacion Femenina Popular (Popular Feminist Organization)
ONIC	Organización Nacional Indígena de Colombia (National Indigenous Organization of Colombia)
ORDEN	Organizacion Democratica Nacionalista (National Democratic Organization)
PCC	Partido Comunista de Colombia (Communist Party of Colombia)
PCN	Proceso de Comunidades Negras
PDA	Polo Democratico Alternativo
PIN	Partido de Integracion Nacional (National Integration Party)
PRI	Partido Revolucionario Institucional (Institutional Revolutionary Party)
PUPSOC	Coordinación Departamental Valle del Cauca del Proceso de Unidad Popular del Sur Occidente Colombiano
SACTA	Supplemental Agreement for Cooperation and TechnicalAssistance in Defense and Security
SIJIN	Seccion de Investigaciónes Judiciales e Inteligencia, de laPolicía (Department of Judicial Investigations and Intelligence)
SINALTRAINAL	Sindicato Nacional de Trabajadores de la Industria de Alimentos (National Union of Food Industry Workers)
SINALTRACEBA	Sindicato Nacional de Trabajadores Cerreceros de Bavaria
SINTRAFEC	Sindicato Nacional de Trabajadores de la Federación Nacional de Cafeteros de Colombia (National Union of Workers of the National Coffee Federation)
SINTRAINAGRO	Sindicato Nacional de Trabajadores de la Industria Agropecuaria (National Union of Agricultural Workers)
SINTRAMIENERGETICA	
	Sindicato Nacional de Trabajadores de la Industria Minera, Petroquimica, Agrocombustibles, and

	Energetica (National Union of Workers in Mining, Oil, Agro-combustibles and Energy)
SINTRAUNICOL	Sindicato de Trabajadores y Empleados Universitarios de Colombia (Union of Colombian University Employees)
TCC	Transnational Capitalist Class
TNCs	transnational corporations
TNS	Transnational State
UDA	Ulster Defense Association
UGB	Union de Guerreros Blancos (White Warriors Union)
UNDP	United Nations Development Programme
UNHD	United Nations Human Development
UP	Union Patriotica (Patriotic Union)
USLEAP	US Labor Education in the Americas
UVF	Ulster Volunteer Force
WB	World Bank
WOLA	Washington Office on Latin America
ZRC	Zonas de Reservas Campesinas

Caribbean Sea

LA
GUAJIRA
ATLÁNTICO Barranquilla
Valledupar
Magdalena
Medio CESAR
San Onofre MAGDALENA
El Carmen
de Bolívar
Monteria Tibu
SUCRE
CÓRDOBA BOLÍVAR Ocaña
NORTE DE
SANTANDER
Apartadó
Curvarado Catatumba

PANAMA

ANTIOQUIA SANTANDER ARAUCA
Tame
Medellin
CHOCÓ BOYACÁ CASANARE
CALDAS
RISARALDA CUNDINAMARCA
QUINDÍO VICHADA
Pacific Bogota
Ocean VALLE DEL TOLIMA
CAUCA META
Cali
CAUCA GUAINÍA
HUILA Calamar
NARIÑO GUAVIARE
Miraflores
Puerto
Asis PUTUMAYO CAQUETÁ VAUPÉS

ECUADOR

AMAZONAS

VENEZUELA

BRAZIL

PERU

Legend
International Boundary
Department Boundary
National Capital
Department Capital

0   200 km

Map of Colombia

Acknowledgements

Writing about paramilitarism and capitalism, while being a mom with twin baby boys, is not an easy task, to put it mildly. This work would not have been possible without the help from my mother and my husband. I am grateful for their hard labour and daily sacrifice as well as for my dad's interesting questions and curiosity about my work. I would like to express my deep appreciation for the support and encouragement given to me by Mark Thomas, Lesley Wood, and my *compadre* and dear friend Peter Landstreet. My intellectual inspirations and motivation have always been strengthened through my interactions with David McNally and Hira Singh whom I profoundly admire with all my heart.

I also thank my friends and colleagues Jeff Webber, Todd Gordon, Andrew Lee, Benjamin Cornejo, Marshall Beck, Rangel Ramos, Kate Pendakis, Janine Muller, Anita Gombos, Rita Kanarek, Katharine King and Jan Hill for their respect for my work and continuous encouragement throughout the years.

To all my *compañeros* in Colombia and Latin America: your trust and faith is what matters the most.

Preface

Several years ago, at an academic conference in Calgary, Canada, where I presented a paper on the structure of the Colombian state's coercive apparatus and its paramilitary ally, one Colombian academic responded to my presentation with the remark that the overall theme of my talk seemed to be 'the same old story: the rich against the poor' – a theme which, according to her, was outdated and no longer reflective of the multiplicity of social actors on the Colombian landscape today. This was just one tiny example of the many different tactics used by those in support of the status quo, among them academics and politicians alike, all of which are aimed at discrediting, dismissing, delegitimizing and ultimately silencing those of us who seek political and economic transformation. We have been labelled as anything ranging from ignorant and naive, to old-fashioned Marxists, to guerrilla spokespersons.

But how can we desist from the struggle for social justice when the faces of poverty and the forces that sustain it are horrifying? While Western headlines on Colombia revolve around sensationalistic accounts of 'narco-guerrillas' and FARC hostages, a two-month-old baby inside a precarious rat-infested dwelling in a poor neighbourhood in Cali was eaten alive by rats after his mother had left him in the care of his six-year-old sister, while she was on the street selling candy to earn a living. This story did not even make the local headlines. How can we believe that the past is really past and Colombia has entered a new era of post-conflict when the paramilitary's surmounting debt with humanity is nowhere close to being paid? Today the families of the victims of paramilitary violence are encouraged to be content as long as the assassins of their loved ones reveal where they buried the chopped up corpses. They can either self-induce a general amnesia and be grateful that they are still alive, or witness the naturalization of the most dehumanizing and cruel acts of violence. The rulers of Colombia tell us that paramilitarism no longer exists, even when masses of people continue to be violently displaced to make way for agribusinesses and resource extraction industries, when those seeking to recover the land that was violently expropriated from them are now murdered, when the extermination strategies against labour unionists have not ceased. In fact there is no better opportunity to demonstrate the relevance of Marx's description of capital as 'dripping from head to toe, from every pore, with blood and dirt'. The unchallenged strength of paramilitary organizations reinforces a message of fear for many and a go-ahead for existing and future groups of paramilitary nature to satisfy their ruthless appetite for resources by following confidently the recipe for guaranteed enrichment through the conversion of human blood into capital. It has been eight years since the demobilization of the AUC. This number has a great significance since it is the maximum sentence under the Justice and Peace Law for those who have committed crimes against humanity. They are now done serving their jail time. This book is the product of a mounting indignation at all this impoverishment and injustice. But it is also an act of resistance against something perhaps even equally tragic, but more subtle.

So far the working majority in Colombia have been losing their land, their natural resources, their labour and civil rights, and increasingly their human rights. But today, Colombia is on the brink of losing its critical consciousness and erasing its collective history. Two major threats are a cause of concern here. First, President Santos' administration is trying and quite successfully managing to completely distance itself from the forces that cause violence and human suffering and has declared that it is working to combat these forces. Former President Uribe's explicit *mano dura* ideology has in the past three years been replaced by a discourse in which the state presents itself as socially progressive, on the side of the people, trying to address the needs of the most disadvantaged sectors. In reality, the essence of the free-market paradigm, which consists of privatizing the gains while socializing the costs, continues with full steam leaving behind wreckage and death. Secondly, the capitalist classes of Colombia (especially the narco-bourgeoisie) have nowadays managed somehow to become the object of admiration, attention and recognition, and serve as a role model for many youths of any social class. The very fabric of popular culture and history is increasingly coming under the influence of the narco-narratives – mostly in the form of books and soap-operas which claim to be based on true stories where the main protagonists are drug-traffickers and those implicated with them. Over the past seven years the despicable popularity of these has pushed into obscurity the ordeals of the assassinated union leaders and their families, the Afro-Colombian families displaced by the oil-palm agroindustry, or the artisanal miners of Cauca violently attacked by state forces. The most fascinating and talked about women are those who have converted themselves into playthings of the *patrones* (narcos / paramilitary / cattle-ranchers). They are the 'heroines' whose biographies are sold by book vendors, not the courageous selfless women of the OFP, or the leaders of the victims' movements, or all those who continue to put their lives at risk on behalf of the excluded and the marginalized. The perverse glorification of armed and wealthy men who seek to increase their fortune even further through violence has become the reference point for what constitutes success and one which thousands of youth aspire to reach either by working for a *patron* or by becoming his mistress. The objectification and commodification of womanhood has become an integral part of this system of domination. Who is going to validate the ideas and painful reality of the multitude which constitutes the unattractive but true Colombian experience?

What is urgently needed in Colombia and what constitutes at the same time a formidable challenge is to launch a counter-ideological campaign of conscientization that can revive once again the dialogue about social inequality, injustice, poverty, ignorance, neglect and the loss of dignity. The voice of existing social movements can be strengthened if millions of other voices join in and collectively renounce the forgetfulness and the erasure of history that the powerful are seeking to impose. The writing of this book is an act of resistance to these deceptions and an attempt to keep alive the struggle for justice by revealing the continuities and novelties of the engine of dispossession, repression and dehumanization. By deconstructing the dominant ideological facade and delving into issues of violent dispossession and paramilitary activity in the 'post-paramilitary' era, the writing of this book is also a practice of freedom. Naming and theorizing paramilitarism is an act of empowerment that speaks directly to the pain of the lived reality of millions of Colombians because it gives expression and meaning to their experience. We cannot let the dictates of the privileged, the violators and the expropriators determine the course of Colombian history and condemn millions more to death or a life without dignity. As Paulo Freire believed, to speak a true word is to change the world. The time has come to build a new Colombia, *una Colombia sin patrones*.

1

Introduction:
The Spectre of Paramilitarism

What you have to understand my dear is that business needs security. We do business and we work with businessmen too. All those Leftists are not good for business. They are trouble makers. That's all they do. We try to establish order, do business which also benefits the community and the poor because we improve the roads, the schools, etc. But the problem is that while we work to bring progress to our country, all they do is put stupid ideas into people's heads. While we construct, they destroy. We don't want communists or socialists or terrorists. You see, we work with the state, we don't work for the state, but with the state. The state doesn't have to give us orders and tell us what to do. It's the reality and our interests that dictate what we do. Those people don't realize that it's the wealthy who give jobs to the poor. And, well, even when our profits come from activities that are seen as all bad, at the end we spend them here, we invest here, you know we benefit our economy, there is nothing wrong with that.

Oscar, member of a paramilitary group,
Department of Santander, Interview 2009

This work employs a Marxist political economy perspective to explore the role of violence in processes of capital accumulation, dispossession and the exacerbation of social inequalities. It is my belief that, while multiple forms of and motives for violence are present in Colombian society, it is possible to discern one pervasive and persistent kind of violence capable of reproducing itself that is of central importance to the armed conflict and to any future prospects for peace. It rests upon the fusion of economic and political power, is spearheaded and organized by considerable sections of the Colombian capitalist classes, and is facilitated through the support of various state institutions. The phenomenon of paramilitarism is the very embodiment of this kind of violence. Hence, the central focus of the book is the nexus between the paramilitary, capital (local and foreign), and the state in Colombia.

On the surface Colombia appears to present a paradox. Of all the countries in Latin America, it is the one that has had the fewest military coups and spent the least number of years under military dictatorship in the twentieth century (Zuleta 2005). It has been repeatedly regarded as Latin America's oldest and most stable democracy (Palacio 1991) and even as one of the longest-surviving democracies in the Western hemisphere (Holmes, Gutierrez and Curtin 2008). Yet, in merely seven years (1988–95) under democratic governments, this country witnessed 28,332 political killings, greatly exceeding the number in each of the other South American nations during their periods of military dictatorship in the 1970s and 1980s[1] (Giraldo 1996). Throughout Colombia's history, violence has been a decisive structuring process (Oquist 1980) and has manifested itself in some of the most extreme and inhumane ways, as Wolfgang and Ferracuti (1967) have eloquently described:

> Almost all of the brutal and senseless paraphernalia of slaying known to history have been exhibited in Colombia. The ever-present feeling of menace, fear, and death, the actual visual presentation of mangled bodies and other sadistic manifestations, together with a desire for revenge in those children whose parents or relatives have been victims of violence, all tend to perpetuate a situation which possibly has no equal in contemporary Western Civilization ... But nowhere in the Western world in recent times since the Second World War has senseless brutality, a genocidal pattern, and a non-war pattern of violence been nearly so total as in the Colombian tragedy. (Cited in Oquist 1980: 276–79)

Zuleta describes Colombia as an 'explosive mixture of democracy and dirty war' (2005: 133). In 50 years (1960–2010) there were at least 61,604 cases of forced disappearances[2] (Mechoulan 2011). In the period 1985–2000, four presidential candidates, over 1,200 police officers, half of the Supreme Court justices and 200 journalists and judges were murdered. Between 1996 and 2002, a homicide was committed every 20 minutes and a kidnapping every three hours. In 2002, Colombia's homicide rate exceeded 40 per 100,000 inhabitants, giving it one of the highest homicide rates in the West (Briceno-Leon and Zubillaga 2002). In 2002 alone, 144 politicians and public officials were assassinated, 124 were kidnapped, and more than 600 mayors were threatened with death (Pardo 2000, cited in Holmes, Gutierrez and Curtin 2008). It is not surprising that Colombia has earned an informal reputation for being the most violent country in the Western hemisphere.

Statistically, this country ranks as the world's most dangerous place to be a member of a labour union. On average, over the last 24 years, every three days one unionist has been murdered (USLEAP 2011). Colombia is also among the nations with the largest number of internal refugees (Moloney 2005). According to the Consultancy on Human Rights and Displacement (Consultoria para los Derechos Humanos y el Desplazamiento, or CODHES), 5.5 million people have been forcibly displaced[3] as a result of violence in the last 26 years (CODHES 2012). Over two million have fled the country since 1985 (Holmes, Gutierrez and Curtin 2008).

Who Exactly is the Paramilitary?

Parainstitutional violence[4] – encompassing violence carried out by paramilitaries, death squads, vigilantes and warlords – has been a long-standing tradition in Colombia. The latter is regarded as one of the few countries in the world where such violence has been so prevalent in recent decades and where paramilitary organizations have amassed such a substantial share of territorial control and political power (Jones 2008). Paramilitary[5] groups, with the complicity or direct participation of state forces, have been responsible for the majority of the murders, torture, forced disappearances, forced displacement, and threats against the civilian population. Even conservative (state) sources confirm the magnitude of civilian deaths at the hands of paramilitaries – 14,476 between 1988 and 2003. During President Alvaro Uribe's first term in office (2002–6), 8,582 civilians were murdered or disappeared by the paramilitary and/or state forces (Boletín Virtual 2009). The political party Patriotic Union (Union Patriotica, or UP) had over 3,500 of its members murdered or disappeared by paramilitary groups between the mid 1980s and the early 1990s (Holmes, Gutierrez and Curtin 2008). Colombian Senator Piedad Córdoba stated during her speech addressed to the European Union in September 2010, 'Colombia is a mass grave, it is the largest cemetery of Latin America' (cited in El Tiempo 2010). Her statement alluded to the number of mass graves[6] that had been discovered throughout the country in 2010 where state and paramilitary forces had buried the corpses of their victims.

Colombian paramilitary organizations are armed groups, created and funded by wealthy sectors of society, with military and logistical support provided unofficially by the state. Their principal aim is to eliminate or neutralize individuals or groups that constitute a threat or obstacle to the interests of those with economic and political power. Murder, torture and threats are typically used by paramilitaries to silence social activists, eradicate support for the guerrillas,[7] and displace people from areas of strategic economic or military importance. Criminal activities such as trafficking, theft, extortion, kidnappings and assassinations are often part of their sources of funding. Paramilitary groups were first created in the 1960s as part of US-Colombian counter-insurgency projects (with the support of sectors of the local elite), and began to expand in the 1980s as large portions of the local capitalist classes (including large-scale landowners, agribusinesses, mining enterprises and drug-traffickers), as well as some foreign companies present in Colombia, took on a leading role in the creation of paramilitary organizations in various parts of the country. The 1980s and 1990s witnessed a considerable growth in their financial and military power as well as rapid territorial expansion. Paramilitary bodies, with cooperation from sectors of the armed forces, the police, and justice system institutions, actively sought to exterminate or at least intimidate any person or group that was considered to be potential collaborator or sympathizer of existing guerrilla movements.

With regard to the composition of paramilitary organizations, it is important to note that they comprise two categories of people. The first is the leaders (using the term in a broad sense) – including the founders or those providing the funding, high-level commanders, and those responsible for major decision-making of a military, economic or other nature which has a direct influence on the organization's operations and determines its future course of action. The leaders belong to the economically dominant classes (landowners, cattle-ranchers, agribusiness owners, mining entrepreneurs and drug-traffickers) as well as those with political power (such as mayors, other politicians, and military and police officials). The second category of paramilitary members consists of the rank-and-file combatants and low-level commanders who are paid a salary. These are the people who perform directly the acts of violence, security duties, and sometimes intelligence gathering. They are usually recruited from low-income sectors of society such as: unemployed youth in urban and rural areas (including youth gangs), *sicarios* (hired gunmen); private security companies personnel; low-rank state army and police personnel; other state personnel who have had military training; those who wish to join the paramilitary for personal reasons (such as having been victimized in some way by guerrilla forces); and forcibly recruited individuals (minors and adults often recruited by deception where, for instance, they are told they will be given a job at a farm or a construction site). The use of the term 'paramilitary' throughout this book encompasses both categories – leaders as well as rank-and-file combatants. When statements are made regarding paramilitary decision-making or economic/political power, obviously these refer to the first category.

Since the 1960s, the state's perception of the principal security threat has coincided with the paramilitary's official enemy – the guerrillas. Thus, throughout major military initiatives sponsored by US administrations, such as counter-insurgency campaigns against the threat of Communism (1960–80s), followed by the War on Drugs (1980s–90s), and finally the War on Terror (2001 onwards), there has been a systematic cooperation between state forces and paramilitary organizations. Regardless of their ideological covers, all of these 'wars' have essentially consisted of military operations and legal measures targeting an 'internal enemy'.[8] Jointly, state and paramilitary violence has facilitated processes of capital accumulation by repressing social movements, eliminating political opponents, displacing populations, intimidating journalists and human rights activists, and engaging in social cleansing.[9] Although, officially, paramilitary groups had demobilized by February 2006 after peace negotiations with the government, in reality since then there has been an upsurge in paramilitary violence.

The growth in paramilitary activities and the territorial expansion of such organizations between 1990 and 2005 was in parallel to the onset of neoliberalism in Colombia. Starting in 2002, neoliberal restructuring was especially accelerated under former President Uribe and comprised the privatization of public services and resources, deregulation of the labour market, increasing the presence of foreign enterprises (especially extractive industries), and drastic reduction of spending on social services. President Santos (2010–14) has enthusiastically continued the neoliberal agenda. The detrimental impacts of these market-oriented policies on human development are clearly evident as the precarious existence of millions of people deteriorates and social inequalities widen further. According to the United Nations Development Programme (UNDP) Report from 2011,

Colombia's Human Development Index[10] and life expectancy are among the lowest in South America (higher only than Bolivia and Paraguay). With poverty at 45.5 per cent, Colombia ranks as the country with the second highest percentage of the population living below the national poverty line in all of South America, after Bolivia (UNDP 2011). Thirteen per cent of the population live in extreme poverty[11] (Prensa Latina 2012). Fifty-five per cent of the population considered economically active (which amounts to 23 million people) have an income that is less than the minimum salary (Caracol 2013), and approximately 20 per cent of the population are homeless (DANE 2009). Around 45 per cent of Colombians work in the informal economy unprotected by labour laws, and almost half of the employed people earn an income less than the legal minimum wage (Prensa Latina 2012). Colombia's rural areas, where 31 per cent of the population live, are ridden with problems of food insecurity, malnutrition and hunger. Half of all rural households experience food insecurity and 20 per cent of rural children suffer from chronic malnutrition (Agencia Prensa Rural 2013). Around 2.5 million children between the ages of six and 17 are forced to work (DANE 2009). The average illiteracy rate is 8 per cent but is as high as 22 per cent among indigenous women (Boletín Virtual 2009). By being complicit in forced dispossession, implementing policies that favour agroindustries and large-scale local and foreign mining companies, combined with the absence of reliable public education, housing and health care, the state has allowed problems of homelessness, landlessness, deterioration in nutrition and health and concentration of landownership to become aggravated.

This bleak picture of human development is accompanied by considerable wealth inequalities. Based on the UNDP 2011 Report, Colombia has the highest income Gini coefficient[12] in the Americas, standing at 58.5. Moreover, it is the third most unequal country in the world after the Comoros Islands and Haiti. Landownership inequality is particularly acute. Sixty-eight per cent of landowners (mostly small-scale farmers) own only 5.2 per cent of Colombia's fertile land (Richani 2007) while 67 per cent of the country's land is in the hands of 4 per cent of the population (Bonilla 2013). In the same way that poverty and social inequality have been a steady characteristic of Colombian society, violence enacted by the dominant classes and the state has been a permanent feature of its political landscape. Any social movements that have sought to establish a more egalitarian distribution and control of productive resources have been met with repression through legal, ideological and especially military means. This continues to be the case.

While Colombia has been named the most violent country in the West, it has had the most sustained economic growth of all Latin American nations (Holmes, Gutierrez and Curtin 2008). It is worth noting that since 1985, while 5,445,406 people were forcibly uprooted from their land (CODHES 2012), 90,000 were disappeared, 95,000 were murdered (Semana 2014b), and more than 2,800 labour unionists were assassinated (El Espectador 2012), gross capital formation in Colombia doubled. Net inflows of Foreign Direct Investment (FDI) also reached a record level in 2005, making Colombia the country with the highest FDI in South America and also surpassing Mexico. Interestingly enough, and parallel to these developments, the military expanded from a force of 167,000 at the beginning of the 1990s to 441,000 by 2008. The defence budget steadily increased from 2.2 per cent in 1990 to almost 6 per cent in 2008, representing the largest share of government expenditure (Richani 2010). Human rights violations by state and paramilitary forces have continued to take place in substantial proportions even after the so-called demobilization of the paramilitary in 2006. Between 1 January 2007 and 31 December 2011, referred to by the Colombian government as a 'post-conflict' era, 1,512,405 people were forcibly displaced (CODHES 2012) and 218 unionists were murdered (USLEAP 2011; El Comercio 2012).

The Significance of Paramilitary Violence Beyond Colombia

The importance of understanding the relationship among processes of violence, capital accumulation and a deepening of wealth inequalities extends beyond Colombia. As Sluka (2000) rightly argues, 'There appears to be a direct correlation between the increasing power and wealth of the elite [within and between countries], the steadily increasing gap between rich and poor, and the growth of state terror, perhaps the three most obvious global characteristics of the last quarter of the twentieth century' (cited in Jones 2008: 32). Although the experience of each country is contingent upon its particular demographic, political, economic, environmental and cultural characteristics, the insights we derive from Colombian social processes have considerable relevance for other parts of Latin America. Neoliberal restructuring of the economy, along with an impoverishment of the working majority, the presence of transnational corporations (TNCs), high levels of violence, human rights abuses of civilians by state forces, and the formation of paramilitary-like forces, are features that countries such as Mexico, Guatemala, Haiti, Honduras and Brazil have in common to varying degrees with Colombia (Koonings and Krujit 1999, 2004; Pansters and Castillo 2007; Rozema 2007; Mazzei 2009).

When it comes to international war, Latin America has been one of the most pacifistic regions in the world in the past two centuries (Pereira and Davis 2000). However, the use of violence in the acquisition of land and resources as well as in the confrontation of peasant and worker mobilizations has historically been a typical characteristic of most of the continent.

For purposes of clarity, the term 'peasant' and '*campesino*' are used interchangeably throughout this book to refer to the Latin American small-scale farmer who engages in the production of subsistence or food crops for his or her family's needs and/or for sale at a local market. An exception to this is where small-scale farmers turn to the cultivation of illicit crops, discussed later the book. The size of a small-scale farm in Colombia is no more than 200 hectares. Koonings and Krujit (2004) observe that social and political violence in the region has appeared to be enduring, despite the consolidation of formal (political, electoral) democratic systems. The findings of the UNDP Report of 2011 confirm violence and inequality as the two defining features of the region currently. According to Heraldo Munoz, UNDP Regional Director for Latin America and the Caribbean, Latin America has the highest income inequality and the most violence in the world – the region represents 9 per cent of the world's population but concentrates 27 per cent of the world's homicides (Domingo 2011). Homicide rates[13] have been increasing steadily in Latin America since 1984 (Pearce 2010). By 1998, violence was the leading cause of death in Latin America among people in the 15–44 year age group (Briceno-Leon and Zubillaga 2002). In fact, today Latin America is second only to South Africa in levels of homicide in the world (Pearce 2010).

While Latin Americans and many people in other parts of the world are aware of the high prevalence of violence in this region, it is mostly the sensationalist simplistic accounts presented by mainstream media that the public is exposed to. The systematic violence employed by the dominant classes and the state's coercive apparatus remains largely neglected in favour of isolated criminal acts in large cities. Numerous scholars have pointed out that today most Latin American societies are primarily urban. It is believed that people move to the cities because there are more opportunities, which is in turn presumed to be sufficient ground for disregarding the question of rural class structure and land ownership altogether. Consequently, the focus on the 'urban' and the 'criminal' obscures the fact that a considerable part of the rural-to-urban migration is actually an intentional product of displacement (whether direct/forced or indirect). This often entails an element of violent dispossession carried out by irregular armed groups and/or state armed forces on behalf of local capitalist sectors and foreign enterprises. The focus on criminal activities in urban centres also misses the violence targeted at popular movements and social actors such as labour unions, women's organizations, and human rights activists, even when these in fact occur in an urban environment. When reported on, such cases are presented as isolated incidents rather than as part of a strategy of repression and intimidation against those who in some way challenge or represent an obstacle to the interests of the dominant classes. Examples of violence associated with land appropriation, displacement and ownership as well as the repression of rural and urban workers' struggles are abundant across Latin America. Yet their coverage is largely limited to journalistic accounts, which have been exposing the growing importance of land in recent years. According to a 2011 report by the UN's Food and Agriculture Organization (FAO), the skyrocketing of food prices has been accompanied by increasing acquisition of fertile land by multinational corporations (MNCs), a process which has been referred to as land-grabbing[14] (Albinana 2012). Millions of hectares of farmland in Latin America have been acquired by corporations investing in the production of food crops and agro-fuels for export (GRAIN 2010).[15] Land-grabbing involves forced evictions, dispossession, migration and the criminalization of those who resist giving up their land (GRAIN 2011).

It is not only foreign investors who are behind the dispossession currently taking place in Latin America. Historically and up to the present, the political and economic power of local elites has been derived from landownership, evident in the very unequal patterns of land distribution. For example, in Brazil, 3.5 per cent of landowners have nearly 60 per cent of the best farmland, while the poorest 40 per cent of farmers have access to merely 1 per cent of the land. Landowners and logging companies continue to force more people off their land, while their private armed forces silence land reform, human rights, trade union and environmental activists. In a period of 16 years (1985–2001) 1,237 murders linked to land disputes were reported in Brazil according to official sources (Frayssinet 2007). This number does not include all the other human rights violations that do not necessarily result in death. In Mexico, the militarization of some regions as part of the War on Drugs[16] has facilitated the operations of extractive industries, as rural activists who organize resistance against mining companies are targeted by state as well as private security forces. For instance, Dante Valdez, who engaged in activism against Minefinders – a Vancouver-based company that operates an open-pit gold mine (Paley 2011) – was murdered by a group of 30 armed men in Madera. Similarly, in Peru mining companies hire private security firms made up of former police and military personnel to target communities who protest the negative environmental and social impacts of large-scale mining (Ford 2009). In Venezuela, former President Hugo Chávez introduced a law in November 2001 aimed at land redistribution for the benefit of poor farmers. The refusal of large landowners to obey the law was symbolized by their act of publically burning copies of the law and broadcasting this on television. From the introduction of the law until 2011, peasants faced a campaign of intimidation and violence enacted by the private armed forces of the landed elite, resulting in the death of 255 people (Fuentes 2011).

Paramilitary-like formations are not limited to agrarian conflicts throughout Latin America. Such forces have also been employed to conduct social cleansing and confront urban criminal gangs in Central America and Brazil. For instance, across various cities in Brazil, what Pearce (2010) refers to as 'para-state death squads' engage in extrajudicial executions of youth gang members as well as other residents in poor communities.[17]

Another reason why the Colombian case bears relevance to other parts of Latin America has to do with the current militarization as well as decentralization and privatization of violence underway in Central America, promoted and organized by the US, the Inter-American Development Bank (IDB), and former Colombian President Alvaro Uribe. In 2010 the US created the Central American Regional Security Initiative (CARSI), with a budget of $165 million and the participation of private security contractors, the CIA, as well as the US and Colombian military forces. The main role in CARSI so far has been played by Uribe, who has been promoting, through a series of conferences[18] across Central America, the Colombian model of decentralized policing through 'public-private partnerships' (that is, cooperation between state military forces and private security contractors) and the expansion of electronic surveillance. Colombian paramilitarism serves as the blueprint for the design of these police and militarization reforms, which are already underway in Honduras,[19] Guatemala and El Salvador (Bird 2011). Such programmes are couched in state security discourses revolving around narco-trafficking[20] and organized crime, nonetheless they create conditions conducive to the emergence of paramilitary structures which would in turn serve to repress social protests and more specifically, as Bird (2011) observes, communities who contest the concessions given to companies involved in palm oil production, petroleum extraction and hydroelectric dams.

Even before CARSI was born, Colombian paramilitarism was already being exported abroad. One example of this can be found in Honduras. In August 2009, two months after the military coup, around 40 former United Self-Defence Forces of Colombia (Autodefensas Unidas de Colombia, or AUC) fighters in Colombia were recruited, with the help of security men working for the paramilitary Eduardo Cifuentes, alias El Aguila, to work for Honduran businessmen and landowners. The monthly salary offered in Honduras was the equivalent of 1.5 million Colombian pesos, plus food and accommodation (El Tiempo 2009). Sectors of the Honduran elite have imported and contracted Colombian paramilitaries to provide security on large estates, sugar cane and African palm oil plantations, to combat urban criminal gangs known as the *maras*, and to back the military coup that removed President Zelaya from office by attacking Zelaya's supporters (El Heraldo 2009). Another example of Colombian para-militarism's extensions abroad comes from Venezuela and is thoroughly documented in Golinger's (forthcoming) work. On 9 May 2004, more than 100 Colombian paramilitaries were captured on a farm outside of the country's capital city, Caracas, by the Venezuelan authorities. They had been planning an attack on the Miraflores presidential palace with the goal of assassinating President Chávez. Years later, former paramilitary commander Salvatore Mancuso admitted in an interview he gave from prison in the US that some Venezuelan military officers had requested his collaboration to execute a second coup d'état against Chávez after the failed attempt in April 2002. Mancuso also revealed that members of Venezuela's anti-Chávez opposition had met with the leader of the AUC, Carlos Castaño, to plan Chávez's demise (Golinger, forthcoming).

In an article on politically motivated violence in South Africa, Duncan (2013) poses the question of whether Colombia's present is South Africa's future, in reference to the growing violence against social movement activists and the informalization of repression through a collusion between the state apparatus and hit squads. Paramilitarism, then, is fundamentally important to the economic and political dynamics of Latin America and possibly other parts of the world. It is a phenomenon that is not likely to disappear any time soon but, on the contrary, is expanding throughout the region as a result of transnational, state-led as well as locally based private initiatives.

Why is it Necessary to Theorize Paramilitarism?

According to the Vice-President of the International Sociological Association, Raquel Sosa, the crucial challenge of the twenty-first century is the confrontation and eventual elimination of the processes of structural inequality that affect millions of human beings today – 'the accumulation and overlap of all types of injustice not only complicates or impedes the realization of legitimate aspirations and rights of the inhabitants of the earth to live in a dignified manner, it also condemns thousands of defenseless human beings to death' (Sosa Elízaga 2013: 1). The intensification of social inequality and its destructive consequences brought about by the economic restructuring done in the name of freeing capital from fetters has now been recognized by sociologists as a current pressing issue. Along these lines, Robinson argues that the relative power of the exploiting classes over the exploited classes has been enhanced many times over in the late twentieth and early twenty-first centuries. However, even more importantly, in his lucid theory of global capitalism he remarks that the latter 'requires an apparatus of direct coercion to open up zones that may fall under renegade control, to impose order, and to repress rebellion ... There must be political authority with a coercive capacity to attempt to secure the environment necessary to undertake accumulation' (Robinson 2004: 137–8). The following imperative question then arises. How is this coercive function organized and performed in the context of a) the increasing difficulty for states to maintain legitimacy in the face of growing discontent with neoliberalism among the labouring classes and b) the pressure to adhere to internationally accepted democratic and human rights principles? The answer to this question lies in the decentralization and privatization of violence and intelligence gathering that has proven to be a ripe ground for parainstitutional formations and especially paramilitary bodies. Through this work, my goal is to advance theoretically significant propositions regarding the role of violence in processes of capital accumulation and primitive accumulation as embodied by paramilitary forces in the era of neoliberal globalization.

For a while, the predominant focus of the literature on violence in Latin America was either on counter-insurgency operations carried out by state military forces or on revolutionary violence aimed at overthrowing the state. In the past 15 years a considerable number of scholars have claimed that violence in Latin America is no longer political but rather criminal and delinquent (Pearce 2010). So, criminal gangs such as the *maras* of Central America and the *favela* (slum) gangs in Brazil, typically comprised of poor urban male youth, have become the object of interest. Political violence is typically conceptualized as either organized by the state or by anti-state movements. Any violence carried out by actors external to the state, who are not guerrillas, is perceived as unrelated to the state and void of any political motives. Thus, all non-state and non-guerrilla violence tends to be lumped into one category and labelled as criminal, showing little awareness for the need to conceptually distinguish paramilitary violence from criminal violence. Holmes, Gutierrez and Curtin write:

> instead of being a deviant case, in many ways the Colombian example is representative of the study of political violence, which tends to be disjoined, conceptually muddled, handicapped by a lack of data, and troubled by political connotations in the definitions given for different types of violence and the relationships among them. (2008: 6)

The role of parainstitutional formations in recent Latin American politics and social conflicts is both notable and notorious (Jones 2008). The development of paramilitarism in Colombia has been crucial to contemporary processes of capital and class formation, particularly in relation to land-grabbing, the operations of extractive industries, and the repression of struggles against privatization and austerity reforms. Most accounts of paramilitary groups, however, are produced by human rights organizations and investigative journalists. Scholarship dealing specifically with paramilitary forces in Colombia or any other country primarily takes the form of case studies limited to a particular time period and/or geographic area. There are neither works that capture comprehensively the mechanisms, strategies and logic of paramilitarism, and explain its reproduction and evolution over time, nor a well formulated and coherent theoretical analysis of this phenomenon.

The need for a deeper inquiry into the Colombian case becomes even more pronounced once we consider the multiple links that exist between this South American country and North America, including migration, solidarity initiatives, the Colombian imports consumed in the North (such as bananas, coffee and flowers – often produced in conditions that violate fundamental labour rights), free trade agreements, and the US and Canada-based corporations operating in Colombia. Especially over the last decade, both Canada and the US have been deepening their economic and (in the case of the US) military ties with Colombia.

According to Gordon (2010) Colombia has been a focal point of Canadian economic and political foreign policy. Canadian companies now hold the dominant share of the exploration carried out by large enterprises in Colombia. Gordon explains that Canada's strengthening of relations with Colombia in the last few years has been due to the latter's abundance of natural resources, aggressive promotion of and support for foreign investment, market-friendly policies and a strong repressive apparatus committed to silencing social movements and other challenges to capitalist ventures. Canadian Prime Minister Stephen Harper views Colombia as an ally of North America in the face of anti-neoliberal governments and movements in the region such as those in Venezuela, Ecuador and Bolivia (Gordon 2010).

Many Canadian corporations in the fields of mining and hydroelectric power operating in Colombia, such as Cosigo Resources Ltd,[21] Embridge,[22] and Conquistador Goldmines,[23] have impacted negatively the living conditions of rural communities – particularly due to the displacement and human rights violations carried out by state forces and paramilitary groups cooperating with the foreign companies (Gordon 2008). Canada's free trade agreement with Colombia[24] is a further step towards consolidating its access to Colombia's markets and natural resources. Of course, in the case of the US, in addition to all the kinds of connections mentioned above, including a free trade agreement,[25] there has been a continuous military intervention, assistance and presence in Colombia that has played a critical role in the aggravation of the human rights situation.

Notwithstanding the magnitude of human rights violations in Colombia and the well-documented relationship between paramilitary forces and the state, various important North American political figures expressed their full support for former President Uribe's administration. In January 2009, outgoing US President George W. Bush awarded Uribe the Medal of Freedom, the highest US civilian award (HRW 2009). On 10 June 2009, after a meeting with the Colombian President in Ottawa, Canadian Prime Minister Stephen Harper stated: 'President Uribe and his government have made very important progress toward sustained peace, security and protection of human rights in their country' (Canada 2009) – all this while 503 unionists had been assassinated from the time Uribe became President (August 2002) up to the point Harper made this statement (ENS 2010). The friendly relationship and support for the Colombian government has continued under President Santos.

Human rights violations carried out by the paramilitary are explained by a defence analyst from the US Strategic Studies Institute in the following way: 'The atrocities of the paramilitaries are not acts of abnormal men, but rather the acts of normal men subjected to and victimized by unremitted violence, who see the disappearance of the guerrillas as the only sure solution to their plight' (Spencer 2001: 2). This author is not alone in producing such a distorted explanation of paramilitary violence. There is a serious need for academics, politicians and policy-makers to have a balanced and comprehensive knowledge of the current forces that fuel the armed conflict in this South American country in order to avoid contributing to the further victimization of its civilian population and of popular movements seeking social justice and democratization.

The bifurcation into political and economic within the academic literature (Pearce 2010; Holmes, Gutierrez and Curtin 2008) inhibits a proper understanding of the nexus between violence and the economy. As a sociological work, this book employs a relational/dialectical approach that challenges and transcends the disciplinary boundary between the economic and political fields of academic inquiry. A relational/dialectical approach views the political and the economic as two dimensions of one social whole where neither has an independent existence and each can only be understood in relation to the other. Thus, I conceptualize capitalist development and violence as mutually determining by examining concurrently the economic underpinnings of violence and the integral part that violence has played in the capitalist development of the country. Such an analysis allows us to grasp better the intersection of conflict and development and, more specifically, in the case of Colombia, paramilitarism and neoliberalism.

I believe that it is not only necessary to acknowledge the relationship between capital and violence but also to reconsider the way it has been conceived. Despite the widespread belief that land is an issue of the past, violence today continues to be largely linked to patterns of land ownership, control and use. While the relationship between violence and capital accumulation has been historical, today there is a novelty in its expression – coercive power is strengthened through paramilitary groups and economic power is strengthened through particularly lucrative activities such as mining, agroindustry and drug-trafficking. For instance, the process of enclosure of territorial mineral endowments and the state's enforcement of private mineral extraction rights is a highly politicized and extremely violent process whose goal is to further the interests of capitalists, as O'Connor and Bohorquez (2010) point out. The relational approach I take allows us to move beyond simply denouncing human rights violations towards a critical analysis that does not treat the various forms of violence as isolated, temporary or accidental occurrences, but rather as expressions of trends that must be considered in relation to larger systems of accumulation.

While there have been some works (mostly by North American scholars) that have critically analyzed the Colombian politico-economic model and have considered the relationship between capital and violence, their focus has been on the imperialist-led militarization of Colombia, the central goal of which is to guarantee security for the operations of foreign enterprises and the implementation of neoliberal reforms that benefit the latter. However, it must be remembered that Colombia has a strong local capitalist class whose wealth is based on land ownership, cattle-ranching, agribusiness (such as African palm oil), mining and drug-trafficking. As O'Connor and Bohorquez (2010) comment, although the neoliberal transformation of Colombia's mining sector is a classic case of the capitalist expansion of the North through accumulation by dispossession in the South, it is no less true that the Colombian ruling class instigated the internationalization of mining over and against popular resistance from artisanal producers, rural social movements, and labour and revolutionary groups. As we can see, locally led and foreign-led capital accumulation, facilitated by neoliberal restructuring, occur side by side and are mutually beneficial most of the time. For instance, the Colombian elite liberalized the economy, thus expanding the opportunities for foreign-led capital accumulation in return for financial and military assistance for its state's coercive apparatus (often channelled into paramilitary groups) which has helped to solidify their power in the face of resistance and revolutionary movements.

My research explores the way in which the local elite employs violence to protect and advance its interests while aggravating social inequalities and giving rise to complex and elaborate armed organizations (that is, paramilitary groups).[26] Based on this focus I hope to contribute to the academic debate on primitive accumulation by highlighting how present-day locally led forms of primitive accumulation operate within a developing country that prides itself as a long-standing democracy. At the same time, I take into account the manner in which local power dynamics interact with the global advance of neoliberalism.

Paramilitarism has come to play an increasingly important role in contemporary armed conflicts as well as in violent social environments in different parts of the world, but it has nevertheless remained largely understudied. It inhabits an ambiguous space between civil society and the state, thus challenging the boundaries between what we would usually regard as military versus civilian. Their multi-faceted nature, destructive capacity, relationship to the state and elite, and the fact that for half a century they have been one of the principal actors in the Colombian conflict, altogether make paramilitary forces an intriguing object of study. The existence of paramilitarism is one of the most important mechanisms that make it possible for a country to maintain a reputation of being a long-standing democracy even though repression, terror and armed force are regularly employed against its civilians. Should these groups be characterized as self-defence patrols, private security associations, criminal organizations, entrepreneurs of violence, warlords, private armies, state-sponsored death squads, terrorist groups or an armed political movement? The way paramilitarism is defined is not solely a matter of academic debate. It can have some very real implications with regard to 1) extending or eliminating the conditions conducive to further human rights violations; and 2) bringing violators to justice. My research demonstrates not only the factors that led to the emergence of paramilitary groups but also the functions they serve and the social sectors that benefit by their existence.

In response to the absence of a comprehensive conceptualization of paramilitary organizations in the academic literature and in the face of considerable distortion in the Colombian and North American mainstream media, I propose a new analytical framework for: 1) understanding paramilitarism in Colombia and the dynamics and evolution of its relationship with the state; and 2) situating paramilitary bodies within the wider context of globalization in relation to producing, maintaining and/or restoring the conditions for capitalist production and accumulation. This framework is grounded in the Colombian case and can serve as a starting point for discerning key elements in the foundations and structures of paramilitarism in Latin America and beyond. Crucial to understanding the social aspects of human experience in our contemporary world is recognizing the global dimensions of economic, political and cultural processes, all of which are embedded in social relations. Thinking globally requires an ability to grasp the ways in which transnational forces interact with local conditions and actors, as they are being facilitated, sustained or contested by the latter. Paramilitarism not only has transnational dimensions, it has also played a crucial role in the transformation of the Colombian state and its incorporation into what Robinson (2004) calls the Transnational State (TNS).

Organization of the Book

Chapter 2 reviews major works from Marx and some contemporary Marxist scholars, with the aim of defining and explaining capital-labour relations, the state and violence in capitalist society as well as in the context of capitalist globalization over the last 30 to 40 years. The chapter then looks at the notion of parainstitutional violence and the ways in which the term 'paramilitary' has been defined and used in different parts of the world. Lastly, I critique existing conceptualizations of Colombian paramilitarism.

Chapter 3 traces crucial patterns of class formation and primitive accumulation throughout Colombia's history from the time of the Spanish Conquest up to neoliberalism. Particular attention is given to key political and economic processes that have laid the groundwork for the birth of paramilitary organizations. The evolution of the state's coercive apparatus, the birth of the guerrilla movements, the formation of paramilitary organizations and their relationship to the state are all discussed in this chapter.

Chapter 4 demonstrates the continuation and in some cases aggravation of human rights violations that have been carried out by present paramilitary forces during the eight years following the demobilization of the United Self-Defence of Colombia (Autodefensas Unidas de Colombia, or AUC) between 2006 and 2014 – a period that has been regarded by government officials as 'post-paramilitary'. Political and media discourses claiming that paramilitarism in Colombia has been eradicated are contested with data illustrating that the military and economic structures of paramilitary organizations remain intact. The chapter elucidates why illegal non-guerrilla armed groups today are of a paramilitary nature and are not merely criminal gangs as the Colombian state would characterize them.

Chapter 5 lays out and explains the guiding principles of the analytical framework I propose for conceptualizing the phenomenon of paramilitarism in Colombia. This framework allows me to shed light on some common misconceptions and frequently asked questions with regard to the relationship between the paramilitary and drug-trafficking, the difference between paramilitary and cartel violence, the relevance of the guerrilla to all this, and the prospects for peace. Chapter 6 presents the conclusions and discusses the broader implications of this study.

2

Foundations for Theorizing Paramilitarism

The central focus of this work is on the role of violence, exercised on behalf of the dominant classes by the state or private armed groups, in processes of capital accumulation.[1] The first half of the chapter discusses concepts from Karl Marx's theory of capitalism and the state as well as the work of several contemporary Marxist scholars relevant to this subject. The chapter then turns to an examination of various contemporary forms of violence exercised for the purpose of advancing capitalist interests by employing the category of parainstitutional armed actors, one of which is the paramilitary. Finally I analyze and critique existing conceptualizations of Colombian paramilitarism.

Capital and Labour

My own interpretation of Marx's work is closest to the theoretical standpoint of political Marxism proposed by Ellen Wood. In her article 'The Separation of the Economic and the Political' (Wood 1981), she proposes an alternative to existing Marxist theory in an attempt to overcome the false dichotomy between the economic and political fields which causes 'some Marxists to accuse others of abandoning the "field of economic realities" when they concern themselves with the political and social factors that constitute relations of production and exploitation' (1981: 12). In my view, coercion and violence have typically been perceived as emanating from the state. The state, in turn, has been conceptualized as belonging to a sphere of its own, separate from society and the economy. Consequently, coercion and violence have been commonly treated as unrelated to economic processes. I, on the other hand, am interested in exploring state and state-sanctioned violence with regard to the function they serve within the terrain of capitalist class domination. As the rest of the chapters in this book will illustrate, in Colombia, violence for the purpose of capital accumulation (through dispossession and repression) is not exclusively exercised by the state but also by paramilitary groups. Even so, the state continues to be the primary agent that makes this possible. This is why the approach of political Marxism is particularly helpful for understanding the relationship between class, violence and the state under capitalism.

Marx: Capital as a Social Relation

Marx provides us with a theory of capitalism that is grounded in people's real lived experiences and oriented towards social action. His work rejects the way classical political economy takes the categories that characterize the capitalist mode of production and presents them as abstract, eternal, universal and natural, thus emptying them of their social content. He detaches us from the normality that has come to surround the various elements of the capitalist structure such as commodities, exchange and labour, and reveals the social relations they conceal. In *The Grundrisse* (1941/1978) and *Capital* (1867/1990) Marx exposes the social bases of the capitalist mode of production by placing social relations at the centre of the analysis: 'The object before us, to begin with, material production. Individuals producing in society – hence socially determined individual production – is, of course, the point of departure' (Marx 1941/1978: 222). The subjection of labour to capital, where the worker has no other way to survive other than by selling his or her labour-power, constitutes the essence of capitalist production and the key characteristic of social relations under capitalism. As Marx explains in *Capital*, capitalism could have never emerged on the basis of merchant capital (buying from one market and selling in another) and interest-bearing capital alone. It requires

the confrontation of, and the contact between two very different kinds of commodity owners; on the one hand the owners of money, means of production, means of subsistence, who are eager to valorize the sum of values they have appropriated by buying the labour-power of others; on the other hand, free workers, the sellers of their own labour-power, and therefore the sellers of labour. Free workers, in the double sense that they neither form part of the means of production themselves, as would be the case with slaves, serfs, etc., nor do they own the means of production, as would be the case with self-employed peasant proprietors. The free workers are therefore free from, unencumbered by, any means of production of their own. (Marx 1867/1990: 874)

Even though the concept of class has been subjected to numerous criticisms, the reality of class remains today.[2] As Miliband argues, the 'Working class across countries remains a distinct and specific social formation by virtue of a combination of characteristics which affect its members in comparison with the members of other classes; the most obvious characteristic is that they get least of what there is to get and have worked the hardest for it' (1973: 16). Marx's concept of class is an invaluable tool for understanding social relations under capitalism. To discuss class we need to first comprehend the connection between the production process and social relations. In other words, it is necessary to grasp the social bases of the production process and, thus, the relations of production.[3] Marx clearly lays out this connection in his lecture 'Wage Labour and Capital'.[4] To begin with, people must modify nature in order to make it useful for their survival. They do so by entering into relations with each other. Production is therefore social and the relations that arise from it are interwoven into the very fabric of society. 'The relations of production in their totality constitute what are called the social relations' (Marx 1847/1978: 207).

Class inevitably rests upon exploitation since in order for a capitalist class (the owners of capital) to exist, surplus value must be produced by workers and appropriated by the capitalist. The way in which capital is produced, or the 'secret of profit-making', consists in the production of surplus value, which is the difference in the value of necessary labour versus surplus labour.[5] As Wood (1981) observes, the category of capital has no meaning apart from its social determinations. Therefore, to understand capital we must first and foremost acknowledge that it is a social relation based on exploitation. As will be illustrated in the subsequent chapters, paramilitarism is an important component of a social structure that rests upon exploitation since it is a strategy that facilitates or enables an increase in the extraction of surplus value in two ways: 1) first and foremost by supplying the principal prerequisite for the extraction of surplus value – the availability of workers who have no other way to survive but to sell their labour – accomplished through the dispossession of people from their means of subsistence; and 2) by repressing workers' discontent over the terms of labour, thus enabling capitalists to increase the rate of exploitation.

A crucial point to take away here is that just as capital is a social relation, so too is class. Class analysis as an explanatory method is still powerful today. The concept of class is useful to understanding the global neoliberal economy today as long as we think of class as a fluid relation of inequality between those who own and control resources and means of production and those who do not. The essence of this relation of inequality is in Marx's words characterized by 'the poverty of the great majority who despite all their labour, have up to now nothing to sell but themselves, and the wealth of the few that increases constantly although they have long ceased to work' (Marx 1867/1990: 873). In the Colombian case, as is demonstrated in the following chapters, such a relation of class inequality can be discerned in different contexts (urban, rural, legal, illegal) and therefore can have different expressions (for example, landowners versus rural wage labourers; mining enterprises versus miners; cocaine laboratory owners versus workers; brothel owners versus the sex workers). We should not conceptualize class only as a fixed category with a specific content (for example, factory owners versus factory workers) since this would render the concept useless by neglecting the form in favour of its content.

Class Relations under Global Capitalism

Since its inception capitalism has had a globalizing tendency which is well captured throughout Marx's work, most famously in *The Communist Manifesto*: 'The need for a constantly expanding market for its products chases the bourgeoisie over the whole surface of the globe' and 'The bourgeoisie ... draws all, even the most barbarian, nations into civilization ... It compels all nations, on pain of extinction, to adopt the bourgeois mode of production' (Marx and Engels 1848/1987: 24–5). Yet many scholars agree that the last 30–40 years have witnessed a qualitatively new stage of world capitalism. One scholar within the Marxist tradition who has produced some impressive accounts of the features and mechanisms of capitalist globalization is William Robinson.[6] He sees globalization as the underlying structural dynamic that drives social, political, economic and cultural processes around the world. For the past 500 years, according to Robinson (2004), capitalism has been an expansionary system with two dimensions: intensive (commodifying public and community spheres that formerly lay outside the logic of capital accumulation; that is, the commodification of social relations) and extensive (expanding outward to new geographic areas that were previously outside the system of commodity production). Today we experience the core of globalization, which Robinson describes as the near culmination of the 500-year process of the spread of capitalism around the world in its extensive and intensive dimensions. This last epoch of global capitalism dates back to the world economic crisis of the 1970s and began to take shape in the 1980s when the prospects for capital to make profits were restored. He lists four key developments that characterize this new epoch:

1) a new capital-labour relation based on the deregulation and flexibilization of the labour-force;
2) new rounds of extensive and intensive expansion, including the vast acceleration of primitive accumulation worldwide as 'millions have been wrenched from the means of production, proletarianized, and thrown into a global labour market shaped by transnational capital' (2004: 24);
3) the creation of a global legal and regulatory structure to facilitate globalized circuits of accumulation (for example, the World Trade Organization); and
4) the imposition of the neoliberal model on countries throughout the world.

Neoliberalism is a system of policies that promotes the interests of private enterprises by eliminating any remaining barriers to capital's search for resources, labour and markets. The essential components of neoliberalism are trade liberalization, privatization, deregulation and austerity. Trade liberalization refers to the removal of any trade barriers, such as tariffs and quotas. Privatization requires the sale of public enterprises and assets to private owners. Deregulation constitutes the removal of government restrictions and interventions on capital to allow market forces to act as a self-regulating mechanism. This process can take on the form of labour or financial deregulation for example. Austerity entails the drastic reduction or elimination of budget expenditures for social programmes and services (Weaver 2000). Neoliberalism can also be described as an ideology, a culture, and a philosophical worldview that takes classical liberalism and individualism to an extreme (Robinson 2004). One feature of the new class relations in this age of 'savage capitalism', as Robinson calls it, has been the gradual disappearance of the peasantry (small-scale subsistence-oriented farming) and their conversion into wage labour on agribusiness farms, in factories or simply as informal workers. The changes in class relations in this most recent epoch of capitalist globalization are linked to each country's incorporation into the global economy which, according to Robinson (2004), constitutes a transnational process that includes the transnationalization of states and classes.[7]

State Power and Class Domination: The Illusion of the Political-Economic Divide

In the following sections, I explore the essential functions of the state under capitalism as well as the relationship between 1) the state and the capitalist class; 2) the state and the means of violence/coercion; and 3) capital and violence. Throughout his work, Marx is uncompromising about the essential role of the state in the development of capitalism, in securing the conditions for capital accumulation and maintaining the dominance of the bourgeoisie. 'The executive of the modern state is but a committee for managing the common affairs of the whole bourgeoisie' (Marx and Engels 1848/1987: 23). The growing power of the capitalist class has been at all times accompanied by their political advance. It is for this reason that Marx believes that class struggle is ultimately a political struggle, as expressed in *The Communist Manifesto* and *The Poverty of Philosophy*: 'Every class struggle is a political struggle' (Marx 1848/1987: 29). What distinguishes Marx's analysis so radically from classical political economy, as Wood (1981) explains, is that it creates no sharp discontinuities between economic and political spheres since he treats both the sphere of the economy and the sphere of politics as a set of social relations. Wood explains that for Marx, 'the ultimate secret of capitalist production is a political one' (1981: 13) because the powers of the state make possible the expropriation of the direct producer, the maintenance of absolute private property for the capitalist, and his control over production and appropriation.[8]

Wood points out that the political dimension is not extraneous to capitalist relations of production even though on the surface it appears to be so. Her analysis is quite useful in that it goes on to explain the reason why the connection between the state and the capitalist class is diluted and may not appear so obvious at a first glance. The two moments of capitalist exploitation – appropriation and coercion, according to Wood –

are allocated separately to a 'private' appropriating class and a specialized 'public' coercive institution, the state. Nonetheless, it is the coercive force of the state that sustains a private 'economic' power which 'invests capitalist property with an authority to organize production itself, an authority probably unprecedented in its degree of control over productive activity and the human beings who engage in it' (1981: 19). Wood revives the political character of Marx's theoretical project and elucidates with exceptional clarity how capitalist expropriation and exploitation divide the arena of political and economic action and how this structural separation (which scholars have replicated into a conceptual separation) has served as the most effective defence mechanism available to capital. Overcoming the political-economic divide is a crucial element in the analytical framework I propose in Chapter 5 for conceptualizing paramilitarism, particularly for understanding how this phenomenon can have simultaneously an economic, political and military dimension.[9] As I demonstrate later on, the very separation between the economic and the political obscures where power really lies and inhibits a proper understanding of the inter-relatedness among paramilitary forces, the state and the capitalist classes in Colombia.

In line with Marx's view on the role of the state in capitalism, Robinson argues that 'The state is itself a class relation that institutionalizes historical constellations of class forces embedded in social relations of production ... A state is a moment of these class power relations congealed in a set of political institutions' (2004: 189). Thus, the state as a class relation also becomes transnationalized under globalization. The concept of the TNS is revisited in Chapter 5 in light of the discussion on paramilitarism and neoliberalism.

Capitalism and Violence

Marx on Primitive Accumulation

Marx's most extensive account of the relationship between capitalism and violence is found in the section on primitive accumulation in *Capital*, Vol. I. As mentioned earlier, the onset of capitalism was made possible by the interaction between the owners of the means of production and those who had to sell their labour-power in order to survive. Marx critiqued classical political economists, such as Adam Smith and David Ricardo, for not offering an adequate explanation of how these two social classes came about, or, in other words, for discouraging inquiries into the concrete social processes through which capital was formed. They normalized capitalist relations of production by presenting them as economic facts completely disembedded from any power relations. Marx ridicules the notion that capitalists are wealthy because they are diligent and frugal and workers are poor because they are 'lazy rascals' – 'Such insipid childishness is everyday preached to us in the defence of property' (Marx 1867/1990: 873–4). He argues that the departure point of capitalism is the forcible separation of people from their means of subsistence (that is, land). The process that divorces the worker from the ownership of the conditions of his own labour 'is a process which operates two trans-formations, whereby the social means of subsistence and production are turned into capital, and the immediate producers are turned into wage-labourers' (Marx 1867/1990: 874). Once the direct producers have been expropriated (robbed of all other ways of survival), they have only one option: to enter the exploitative relationship offered by those who have now come to monopolize the means of production. 'So-called primitive accumulation, therefore, is nothing else than the historical process of divorcing the producer from the means of production' (Marx 1867/1990: 874–5). Marx highlights the role of violence by demonstrating how the history of this expropriation 'is written in the annals of mankind in letters of blood and fire' (875) and that 'In actual history, it is a notorious fact that conquest, enslavement, robbery, murder, in short, force, play the greatest part' (874). His powerful and detailed historical account of the transition from feudalism to capitalism in England between the 1400s and 1700s illustrates the expropriation of small-scale landholders (including serfs and independent peasants), the appropriation of common lands, and the sell-off of state and Church estates.

The process of primitive accumulation in that historical period consisted of a number of mechanisms: legislation, armed force, the public debt, the international credit system and colonialism. The first two are of particular interest here. After the Civil War of 1642, most of the landed ruling class in England was no longer feudal in nature and there was a drive towards a transformation of what used to be feudal property[10] into commercial farms (capitalist property). Peasants became either employed as agricultural wage labourers, tenants on such farms, or migrated to the towns. Legislation and armed force were two very important instruments employed to expropriate the peasants and subsequently convert them into a proletariat. In the fifteenth and sixteenth centuries, small peasant properties and common lands were usurped through individual acts of violence. Marx rightfully describes capitalism as 'dripping from head to toe, from every pore, with blood and dirt' (1867/1990: 926). By the eighteenth century, the law itself became 'the instrument by which the people's land is stolen'[11] (885). Legislation was a key instrument for the conversion of 'land into a merely commercial commodity, extending the area of large-scale agricultural production, and increasing the supply of free and rightless proletarians' (885).[12]

Legislation and armed force were not only applied during the process of separating the rural population from the land, but also in the transformation of expropriated people into wage earners. We should note here Marx's recognition of the crucial role played by the state in enforcing mechanisms of primitive accumulation and protecting the interests of the bourgeois class in general. In order to address the issue of social order as the number of expropriated increased, legislation was used to criminalize those who did not participate in the capitalist system of production (such as beggars and delinquents), and ordered the use of most horrifying forms of physical punishment. Marx provides a long list of examples of this 'bloody legislation'[13] from the late fifteenth to seventeenth centuries, such as 'whipping and imprisonment for sturdy vagabonds. They are to be tied to the cart-tail and whipped until the blood streams from their bodies' (1867/1990: 899). And 'the ear-clipping and branding of those whom no one was willing to take into service' (901). The use of violence by the state on behalf of the capitalist classes was manifested in the combination of laws and the violence-based mechanisms employed to enforce them. Marx eloquently summarizes it in the following way: 'Thus were the agricultural folk first forcibly expropriated from the soil, driven from their homes, turned into vagabonds, and then whipped, branded and tortured by grotesquely terroristic laws into accepting the discipline necessary for the system of wage-labour' (899).

Marx's account illustrates that the process of primitive accumulation is vital to the creation and survival of the capitalist mode of production since it generates the principal elements necessary for its functioning – the availability of wage labour, natural resources and markets for commodities. The creation of a dependent labour force which had no other means of survival was an indispensable factor in the development of capitalism. Another substantial accomplishment that served as an impetus was the opening of access to natural resources through the change in property rights, particularly land, as the latter now became an absolute property. Furthermore, as modes of self-provisioning disappeared as a result of the eviction of the agrarian population from the land, there was a creation of demand for capitalism's products. Once capitalist relations of production are established, primitive accumulation mechanisms continue to be at work:

> The rising bourgeoisie needs the power of the state, and uses it to 'regulate' wages, that is, to force them into the limits suitable for making a profit, to lengthen the working day, and to keep the worker himself at his normal level of dependence. This is an essential aspect of so-called primitive accumulation. (Marx 1867/1990: 899–900)

According to Marx there were other moments of primitive accumulation that relied on violence – conquest and colonialism in the Americas and India, the enslavement of the indigenous in the Americas, and the slave trade of African people. 'The treasures captured outside Europe by undisguised looting, enslavement and murder flowed back to the mother-country and were turned into capital there' (Marx 1867/1990: 917).

The Theoretical Debate Over Primitive Accumulation

There has been a long-standing debate within Marxism over the meaning and relevance of Marx's concept of primitive accumulation.[14] On one side we find scholars, such as Zarembka (2002), who maintain that primitive accumulation was tied to a particular historical epoch preceding capitalism. In that sense, the essence that characterizes it is a time-specific dimension – a transition between two modes of production. On the other side of the debate, primitive accumulation is seen as defined by its systematic character – the separation of the conditions of production from the labourer – which is constitutive in the form of capital and has to be reconstituted continuously as capital constantly returns to its beginnings (Bonefeld 2001). My use of the concept to examine the role of violence within Colombia's political economy is based on this second interpretation. Capitalist production depends upon the reproduction of the capitalist–wage labour relation. For this relation to exist, in turn, it is necessary to prevent workers' access to the means of subsistence. As Marx clearly states, 'As soon as capitalist production stands on its own feet, it not only maintains this separation, but reproduces it on a constantly extending scale' (Marx 1867/1990: 874). Given that working-class struggles, which represent a refusal to accept capital's requirements as natural laws, are a continuous element of the capitalist relations of production, capital must continuously engage in strategies of primitive accumulation to recreate the basis of accumulation. In other words, the inherent continuity of social conflict within capitalist production therefore implies capital's inherent need for processes of primitive accumulation. As Harvey has put it: 'A general re-evaluation of the continuous role and persistence of the predatory practices of "primitive" or "original" accumulation within the long historical geography of capital accumulation is ... very much in order' (2003: 144).

My work uses Marx's concept of primitive accumulation to examine the Colombian internal armed conflict as a product of intersecting historical relationships – mainly the impoverishment of the greater part of the population and the acts of violence that dispossess and keep people inside exploitative social relations. Discussion of this can be found in Chapters 3 and 4 where I trace how methods of separating workers from the means of production have manifested themselves in Colombia throughout major historical periods and have continued today, particularly in the prole-tarianization of small-scale producers (that is, the creation of landless peasants) and the repression against social movements that oppose the dominant politico-economic model currently in place.

Capitalist Violence Outside the State Apparatus

Political or Criminal?

Violence has been a constant feature of the Latin American political landscape from colonial to contemporary times.[15] As Pearce puts it: 'the history of Latin American nations from their foundational myths to their historical birth is dripping with blood. But violence is not confined to the distant past. From the twentieth to the twenty-first century Latin American nations have experienced every conceivable form of violence' (2010: 287). While the subject of collective violence in Latin America has generated a vast array of studies, it is possible to identify three dominant fields of scholarship according to the violent actors upon which the focus is placed. The first area comprises works on counter-state violence – that enacted by armed organizations seeking to take over state power with the purpose of bringing about a structural transformation of society by constructing new models of economic and political organization, production and distribution of wealth – also known as revolutionary violence.[16] These works were mostly published from the 1960s to the 1990s, reflecting the emergence of various revolutionary or rebel movements in countries such as Nicaragua, El Salvador, Guatemala and Colombia.

The second field of literature focuses on state-sanctioned violence or state terror, carried out by state agents against subversives or those considered to be their sympathizers along with political opponents and any other individuals and organizations perceived to be Left-leaning. It covers the military dictatorships and authoritarian regimes of the 1960s through to the 1990s. The third area of scholarship (1990s onwards) deals with criminal or delinquent violence aimed at material gain (for example, robbing, kidnapping and trafficking illegal commodities).[17] However, the increase of violence alongside democratic transitions across Latin America should not simply be regarded as the result of a growth in criminality (particularly youth gangs and drug-related violence) and dismissed as politically irrelevant. A very recent body of literature, small but growing, has taken up the question of current forms of parainstitutional violence and their links to the state. Such scholars[18] have begun to research the murky present-day violent landscapes in Latin America and have exposed a variety of other illegal armed actors, such as private justice groups, self-defence forces, death squads, and paramilitary groups, none of whom aim to take over state power (that is, they are not guerrillas but nor are they criminal gangs either).

Some scholars (for example, Romero 2000; Spencer 2001; Bejarano and Pizarro 2002; McLean 2002) view the existence of these armed actors as a sign of state failure or state weakness due to the state's loss of monopoly over the means of violence. However, it is important to note that parallel to the rise of such groups, the state's coercive apparatus continues to play an important role in maintaining security and dealing with the 'internal enemy' (that is, individuals and/or organizations who represent a challenge or an obstacle to the economic and political interests of the dominant classes and the state or express non-conformism with the current politico-economic model). It is true that state terror no longer occurs in the same magnitude as it did in the dictatorial era, nonetheless repression, extra-judicial executions, and other types of human rights abuses, as well as an increase in the use of force by the military and the police in general, continue. Robinson (2003) attributes the continued active role of the state's coercive apparatus against the 'internal enemy' in the post-dictatorship era to a crisis of hegemony.[19] In this context, in addition to relying on the ideological apparatus, in order to ensure social order and security the state engages in direct coercion accompanied by the expansion of law and order policies that increase militarization and criminalize dissent. In the same vein, Huggins (1991) points out that Latin American justice systems still contain many structural supports for authoritarianism. Furthermore, as resistance against global capitalism grows, new waves of militarization[20] are underway in the region, including mounting surveillance, policing and militarized social control (Robinson 2004). It is not uncommon for many non-guerrilla illegal armed organizations to sustain mutually beneficial relationships with members of state institutions or to be comprised of active state security personnel.

A question that Pearce poses becomes very important in this context: 'Are we witnessing, in fact, not state failure in Latin America, but a new perverse form of the state?' (2010: 298). Latin American elites (both locally oriented as well as the transnational fractions[21]), today, as they did hundreds of years ago, rely on violence that ultimately protects their interests, and consent to state security acts aggressively targeting those who challenge these interests. The 'perverse state', dedicated to the preservation of elite rule, transmits and reproduces violence by engaging in violent acts, by being complicit in the violent acts of other actors whom it tolerates, and by neglecting to address private expressions of violence.

Why is the state's lack of monopoly over the means of violence not a case of state failure or state weakness? Firstly, because state failure 'does not convey the systematic way in which the Latin American state enables multiple forms of violence to spread across diverse social spaces and the political capital it accumulates as a result' (Pearce 2010: 295). As Pearce eloquently argues, the Latin American state claims its legitimacy not from a monopoly over the means of violence, but from its lack of such a monopoly. The sources of disorder, such as criminals and youth gangs, provide the justification for the state violently imposing order and re-establishing its authority every time. Secondly, with the exception of the guerrilla, many illegal armed groups are not 'counter-state', but operate with more or less the same general goal – to preserve the status quo, suppress dissent, and protect and advance the interests of the wealthy classes. Thus, as Pearce points out, the state does not regard this as an absence or loss of the monopoly of violence since it has never aspired to exercise such a monopoly. On the contrary, it welcomes these alliances.

This is precisely the case with the paramilitary and the state in Colombia, as the next three chapters will illustrate.

The current juncture of state and non-state violence challenges the boundary existing between notions of 'civilian' and 'military'. As Pereira and Davis explain, the literature which assumes that civilian control of militaries is automatically restored under democracy and that subsequently civil and human rights would be guaranteed, plays down the contemporary changes 'in military role definition that preserve or expand the armed forces' autonomy in such areas as drug interdiction, counter-insurgency, and crime fighting ... it ignores the hybridity and complexity of much state coercion and violence' (2000: 7). For all these reasons, it is very important to recognize the peculiarities of this coexistence between the state's coercive apparatus and illegal armed groups and to understand the restructuring of the former in relation to the latter.

Parainstitutional Violence

The concept of parainstitutional violence is quite useful for this purpose. First, let's look at the term 'parainstitutionality', coined by the Colombian scholar German Alfonso Palacio in 1991 and grounded in the Colombian political context. Palacio defines parainstitutionality as 'a series of mechanisms of social regulation and conflict resolution that do not rely on formal constitutional or legal means, but are governed by informal arrangements and ad hoc mechanisms' (1991: 106). These mechanisms are manifestations of, and at the same time a solution to, the need to both respond to social conflict and to secure the conditions for capital accumulation. For Palacio, paramilitary groups, guerrillas, *sicarios* (hired gunmen) and urban protests are all different expressions of parainstitutionality. Notwithstanding this, he recognizes that most parainstitutionality is linked to drug-traffickers. A couple of other Latin American scholars have taken the concept and applied it only in the context of violence by right-wing armed groups. Medina and Tellez (1994) define parainstitutional violence as violence that does not have as an objective the transformation of society, but rather seeks to guarantee, complement and supplement the state's adequate functioning when the latter is not capable of doing so due to its limitations. It is parainstitutional due to the fact that it accepts the objectives of the existing regime. Its ideological base is grounded in the counter-insurgency strategies of the National Security Doctrine. According to these authors, parainstitutional violence is a feature peculiar to the coexistence of formal constitutionalism and electoral democracy on one hand, and the use of force to pursue economic and political interests on the other. It is extra-legal in the sense that it is pursued by state agents outside the boundaries of legality, or by non-state agents. It is also 'uncivil' in the sense that it undermines the constitution of citizenship as a principle based on non-violence and the rule of law.

In sum, the state in most of Latin America, even under democratic regimes, continues to be an important agent of violence albeit not having monopoly over the means of violence. The Marxist view of the state's coercive apparatus as serving to maintain the domination of the capitalist classes continues to be highly relevant to the present-day Latin American reality. At the same time, it is necessary to recognize the diversity of actors outside the state who engage in collective violence today. While the proliferation of urban criminal and youth gang violence is undeniable, this phenomenon alone does not explain the essence of the violent forces that accompany the geographic and qualitative advancement of global neoliberalism. If we want to comprehend the forces that facilitate processes of capital accumulation through repression and dispossession, then we need to look beyond gangs and towards the state, as well as armed groups operating outside the state, but in cooperation with or tolerated by it. To understand the violent side of global capitalism with its local expressions, it is necessary to recognize how the state's coercive apparatus coexists with such groups and shares control over the means of violence with them, and why this is functional to both sides. The following section explores this subject with regard to four types of parainstitutional actors.

The concept of parainstitutionality comes in quite handy in that it helps us identify armed actors outside the state who have a political dimension to them. However, the way the concept has been used so far has been inconsistent since Palacio (1991) intended it to cover both paramilitary and guerrilla groups, while Medina and Tellez (1994) have modified its definition to refer to right-wing armed groups only. Having a concept that equates the guerrilla and the paramilitary is of little value from a politico-economic point of view, since placing these two armed actors in the same category is only possible if they are totally emptied of their social bases, history and objectives. In this sense, I support Medina and Tellez's use of the term. Finally, I believe that any analysis of violence in Latin America today must not be detached from the social context in which it is embedded – this includes the class relations and system of domination it serves to sustain.

The Myriad of Today's Parainstitutional Violent Actors

This work aims to offer a comprehensive analytical framework that enables us to account for the new phase of paramilitarism in Colombia as part of the evolution of a historical and structural systematic reliance on violence by the state and dominant classes to secure capital accumulation. It would therefore be helpful to first provide some background on different parainstitutional forms of violence that have been studied by scholars so far, in order to situate the paramilitary in relation to the state as well as the myriad of illegal armed actors that exist today. Ron's argument about the increasing tendency of states to privatize violence in response to human rights concerns points to the growing importance of paramilitarism as a subject for academic inquiry:

> Although human rights monitors seek to limit state violence to create a better world, they may sometimes simply drive the violence underground. Faced with restrictions on who and how they can kill, state actors may hand the violence over to semiprivate gunmen, hoping these fighters can accomplish what the state was prevented from doing by courts and human rights activists. (Ron 2000: 308)

Defining Paramilitarism

The American Heritage Dictionary defines paramilitary as: 'Of, relating to, or being a group of civilians organized in a military fashion, especially to operate in place of or assist regular army troops.' The term can also be used to refer to a member or members of a paramilitary force (American Heritage 2009). According to the Collins English Dictionary, the term 'paramilitary' means 'denoting or relating to a group of personnel with military structure functioning either as a civil force or in support of military forces' (Collins English 2003). The Cambridge Dictionary defines it as 'a group which is organized like an army but is not official and often not legal; connected with and helping the official armed forces' (Cambridge 2012). The Merriam-Webster Dictionary's definition states: 'of, relating to, being, or characteristic of a force formed on a military pattern especially as a potential auxiliary military force' (Merriam-Webster 2012).

As we can see, the common elements across all these definitions include:

1) a civil force (group of civilians);
2) organized and functioning as a military force;
3) not part of the state's formal armed forces; and
4) operating or having the potential to operate in a way that assists the state's armed forces.

According to Eseverri (1979), the author of the *Dictionary of Etymology of Spanish Hellenisms* (cited in Giraldo 1996), in Spanish, there are three meanings of the preposition 'para': 1) approximation; 2) transposition; and 3) deviation or irregularity. In effect, this preposition is utilized to make reference to something which is next to, adjoining, or similar to, but which at the same time is beyond, outside of, or departing from the entity denoted by the principal body of the word. The concepts of proximity and deformation are integrated in the meaning of this preposition. Thus, 'paramilitarism' denotes activities close to the military, but which at the same time deviate from or are irregular to the militia (Giraldo 1996). The Colombian Commission for the Study of Violence defines paramilitaries as 'those private and/or state-affiliated organizations that use violence and intimidation to target and/or eliminate groups and individuals seen as subversive of the social, political, and economic order' (Jones 2008).

According to Jones, paramilitary organizations are varied and highly mutable institutions across Latin America and worldwide. Kaldor (2007), whose work has been recognized as a core text in security studies, defines paramilitary groups as autonomous groups of armed men generally centred on an individual leader. She positions paramilitary forces in the context of what she calls 'new wars', referring to the fusion of elements of organized crime, war and large-scale human rights violations. Her research places heavy emphasis on identity politics with regard to the armed conflicts in which paramilitary groups emerge. She draws particularly on cases from the post-Cold War era in Eastern Europe and Africa.

Most other works that offer conceptual definitions of paramilitary forces emphasize the cooperation/support aspect in their relationship with the state. Huggins (1991) equates paramilitary groups with death squads (in the context of Latin America) and describes them as 'assassination squads' that exhibit a right-wing ideology and engage in human rights violations in which the state is complicit or involved. Similarly, Torres-Rivas (1999) places emphasis on the function of paramilitary groups and defines them as unofficial security forces that serve a military or quasi-military function. Warren (2000) describes the paramilitary as death squads composed of former members of the army and police which operate with impunity and benefit from direct or indirect logistical institutional support. Mazzei (2009) defines paramilitarism as a strategy that enables states to avoid appearing as direct sponsors of violence while countering struggles that seek reform or social change. Conceptualizations of paramilitarism with regard to the Colombian case specifically are discussed later in this chapter.

Paramilitarism in Different Parts of the World

The purpose of looking at examples of paramilitarism from around the world is to give a sense of the varied meaning of the term 'paramilitary' according to the geographic location and the political, ethnic and economic context in which it has been used.[22] Ron's (2000) study of paramilitary forces in the context of the Bosnian civil war argues that Serbian paramilitaries resembled Latin American death squads in that they were clandestine semi-autonomous state agents using illegal violence (extrajudicial executions, torture and rape) to achieve political goals against a clearly defined target population. However, their operations targeted not citizens of their own country but rather Bosnian Muslims and Croats in Bosnia, particularly in contested territories claimed by Serbian nationalists. The reason for this was that local and international norms prohibited military action beyond Serbia's official borders and prompted Serbian officials to subcontract semi-private groups. The other difference from Latin American death squads was that they targeted their victims for their ethnic identity rather than political affiliation. In sum, these forces started out as extra-territorial agents capable of clandestinely projecting Serbian state power abroad. They were devised by the Serbian state as a way of circumventing territorial restrictions on Serbian state action, and their mission was defined as the defence of the Serbian nation (Ron 2000).

In Northern Ireland, Protestant Loyalist paramilitary groups have been described as pro-state terrorist organizations in support of the Loyalists who want Northern Ireland to remain part of the United Kingdom. The two main groups are the Ulster Volunteer Force (UVF), founded in 1966, and the Ulster Defence Association (UDA), founded in 1971. Hard-line offshoots of these emerged in the 1990s. These organizations, as well as others that have emerged more recently, engage in intimidation, killings and bombings, and their targets are mostly Catholic civilians (Council on Foreign Relations 2005). Another example comes from Hungary, where the Better Future Militia in the town of Gyonguospata has been described by the media as a right-wing paramilitary group that works in cooperation with the right-wing Jobbik Nationalist Party and targets the Roma population (Guardian 2012).

As we can see, the term 'paramilitary' has been used in different political and geographic contexts to refer to a wide spectrum of armed bodies, ranging from large, well-known, heavily armed, long-term organizations operating across a considerable geographic area to small spontaneously formed armed groups confined to a specific location. Nonetheless, one of the steady features across the different cases is the paramilitary groups' pro-state stance – in other words their favourable attitude towards the state or the political party in power – as well as the state's tolerance, support or promotion of these groups. In the examples from Serbia, Northern Ireland and Hungary the victims are identified mainly on the basis of their ethnic and/or religious affiliation[23] and the violence occurs in the context of a conflict between different ethnic groups (often involving different religions). While the conflict may or may not have complex implications with regard to territories, political rule and so on, the 'enemy' of paramilitary groups is perceived as someone culturally different from them and from those whose interests they represent.

Quite different manifestations of paramilitarism are found across Latin America and the Caribbean, where ethnic identity or religion is irrelevant with regard to the conflicts within which these groups emerge as well as the criteria they use for labelling their 'enemy'. In Peru, for instance, paramilitary groups were first born in the 1980s, through the institution of autonomous peasant patrols (*rondas campesinas*). These groups have been described as organic parainstitutional formations that resisted and confronted the imposition of guerrilla power on peasant communities.[24] In the early 1990s, after Alberto Fujimori became President, the state military recognized their potential as a support in the war against the insurgency. Consequently, the *rondas campesinas* were armed professionally and integrated into the state's counter-insurgency strategy.[25] After the Sendero Luminoso was defeated, the *rondas* reverted back to being a neighbourhood watch group (that is, protecting mainly against theft) (Jones 2008).

In Guatemala, paramilitary organizations have played an important role since the beginning of the counter-insurgency war in the 1950s. The Guatemalan state created extra-legal armed groups to combat the guerrillas and track down subversives. Two of the most feared groups were coordinated by the right-wing political party Movement of National Liberation (Movimiento de Liberación Nacional, or MLN) and received financial and logistical support from large-scale landowners and commercial farmers. Throughout the 1960s, more groups were formed jointly by the state military and the rural elite. Their members were given the right to carry unlicensed weapons and the role of serving as the 'eyes and ears of the army' in the fight against the guerrilla. In the 1970s and first half of the 1980s, the Guatemalan state centralized the repressive apparatus and paramilitary groups came to be more directly ruled by the state military (Bastos 2004, cited in Granovsky-Larsen forthcoming). The Civil Defence Patrols (Patrullas de Autodefensa Civil) were instituted by the Guatemalan state in the 1980s for the purpose of consolidating military domination over rural communities and isolating Leftist rebels (Jones 2008). The 1996 Peace Accords ended more than four decades of armed conflict; however, according to Granovsky-Larsen (forthcoming), during the contemporary era of post-war democracy, the historical reliance on privatized violence has continued and paramilitary forces have become vital to the present power of the political and economic elite. He explains that during the last 15 years or so, three categories of armed actors have operated as paramilitary forces.[26] These include: death squads, which have a direct link to the military apparatus and are hired to protect Guatemala's most powerful interests; vigilante groups (many of them legalized since 1999 as Local Citizen Security Groups, or Juntas Locales de Seguridad Ciudadana); and private security guards especially in situations where they act against organized peasant communities.

In El Salvador, the National Democratic Organization (Organización Democratica Nacionalista, or ORDEN) was the primary paramilitary body founded by the military dictatorship in the 1960s. By the mid 1970s, ORDEN had reached close to 100,000 forces some of whom engaged in intelligence gathering and others in counter-insurgency operations against guerrilla groups. Although it was officially disbanded in 1979, subsequent paramilitary groups were created in the 1970s and 1980s by hard-line factions of the elite and the military to protect the dominant classes from reforms that might redistribute wealth (Mazzei 2009).[27]

In Mexico, paramilitary groups have been known to be mostly concentrated in the state of Chiapas.[28] These groups were formed by the joint efforts of the Institutional Revolutionary Party (Partido Revolucionario Institucional, or PRI), wealthy business owners or ranchers, and the military and police, following the birth of the indigenous rebel movement, the Zapatista Army of National Liberation (Ejército Zapatista de Liberación Nacional, or EZLN), in 1994. Mazzei's

(2009) study of paramilitary organizations in Chiapas makes a point of distinguishing them from private armed groups created by landowners only, known as *guardias blancas*. The latter, according to Mazzei, target peasant organizers, land reform advocates and critics of the established order for the direct purpose of protecting their boss (the rancher) and his or her property. The paramilitary groups, on the other hand, target the same type of individuals and/or organizations except that it is not only those who threaten a specific landowner, but rather all such individuals and organizations in general because of the threat they pose to the politico-economic system as a whole. The fact that politicians and the military participate in the creation of paramilitary units enables the latter to be trained, armed and organized with a high degree of professionalism. While the *guardias blancas* are property-oriented security forces, the paramilitary has a broader political purpose:

> Thus, while White Guards [*guardias blancas*] defended property, paramilitaries defend the PRI party dominance that protects the large property holders from land redistribution and assigns a status and power via land ... paramilitaries are not reactive or defensive in nature; they are proactive and organized for the purpose of offensive attacks. (Mazzei 2009: 36)

In Haiti, during the past 50 years various paramilitary groups have emerged, disbanded, been absorbed into police forces, or replaced by new ones. What they have all shared in common is their role in 'crushing the Haitian people's experiment in popular democracy' (Sprague 2012: 11), characterized by their affiliation to varying degrees with dominant national and transnational social groups, state army and police as well as their repression of the popular classes. Sprague argues that thanks to these paramilitary networks, the right-wing in Haiti is in its strongest position in decades.

As we can see, in a number of Latin American countries between the 1960s and the present, paramilitary bodies have been formed through various combinations of initiatives led by the state military and capitalist classes for the purpose of protecting their interests and power. This has been done by supplementing the state in its counter-insurgency offensives and repressing sectors of the civilian population considered current or potential allies of the guerrilla as well as those who seek social transformation through non-violent means. Therefore, the criteria that determines whether a person/organization constitutes an 'enemy' (i.e., is to be victimized by paramilitary forces) is not their ethnic or religious identity but rather their objective social position vis-à-vis capital and/ or their subjective position (or stance with regard to the status quo). Hence, in the Latin American context, paramilitary organizations must be comprehended in relation to the development of capitalism.

Death Squads, Vigilantes, Warlords: Distinguishing Paramilitary Groups from Other Violent Parainstitutional Actors

Since the paramilitary as an armed actor has often been equated to or conflated with death squads, vigilantes and warlords, this section reflects on how the paramilitary is similar to and different from each of these other three types.

Death squads are clandestine and usually irregular organizations, which carry out extrajudicial executions and other violent acts (torture, rape, arson, etc.) against clearly defined individuals or groups of people (Campbell 2000). Those who see the paramilitary as the unofficial right-hand of the state, employed to do its 'dirty jobs', have often equated this armed actor with death squads. It is true that both death squads and paramilitary groups in Latin America are founded upon a counter-insurgency ideology which dictates that the guerrilla as well as any individual, organization, or movement which seeks social change, challenges the current politico-economic model, or advocates human rights, should be viewed as an enemy and targeted by violent means such as murder, torture or disappearance. When we consider the agents responsible for the creation of death squads and paramilitary groups, we see in both cases the direct participation of both the state as well as civilians, which in turn problematizes the boundary between state and civil society. Similarly to death squads, paramilitary groups carry out executions of individuals deemed by the state as enemies, to enable the state to maintain a democratic image through 'plausible deniability'. Many of the tactics used by death squads are also used by the paramilitary – for instance carrying out massacres,[29] selective assassinations, or leaving mutilated corpses in public places.

There are, however, some substantial differences between these two types of parainstitutional armed actors. One central feature of the definition of death squads, which all of the works reviewed earlier mention, is their clandestine nature. Paramilitary organizations on the other hand are not clandestine, since they have names, often wear uniforms, and openly let citizens know about their existence, their demands, and the consequences of not complying with their rules. Paramilitary units commonly establish dominance in a certain area and remain there for a long time, controlling many aspects of the local social life, maintaining relationships with informants, and 'cleaning' the area of subversives. A second key distinction in the Colombian case is that the initial creation of paramilitary groups in the 1960s by the state and with the support of elite sectors of society was followed by the formation of numerous other paramilitary groups where sectors of the capitalist class became the principal leaders, while the state assisted them in various ways. Thus, it was civilians who took on the primary role in the creation of these organizations from the 1980s onwards, unlike death squads where members of the state usually have a leading role in their formation and operations. Thirdly, the main job of death squads is eliminating people through extra-judicial executions and disappearances. Paramilitary groups, on the other hand, engage in a very wide range of activities such as forced displacement, extortion, intelligence gathering, social cleansing and security provision. Another fundamental difference is that while death squads have links to the state military, paramilitary groups not only have links to the military but have also penetrated profoundly many other major state institutions – such as the justice system (where by destroying evidence and manipulating witnesses state authorities ensure impunity for paramilitary crimes), or the Colombian Institute for Rural Development (Instituto Colombiano de Desarollo Rural, or INCODER)[30] where officials provide paramilitaries with legal titles for land which has been illegally and forcibly appropriated.

With regard to vigilantes, Rosenbaum defines them as engaging in 'acts or threats of coercion in violation of the formal boundaries of an established sociopolitical order which, however, are intended by the violators to defend that order from some form of subversion' (1974: 542). One of the activities of paramilitary groups is social cleansing, and it is in this respect that they resemble vigilantes. As part of their attempt to dominate a certain area, it is common for paramilitaries to distribute flyers with warnings to anyone perceived as a disturbance to the social order, including thieves, beggars, street prostitutes, the poor, the mentally ill and drug-addicts. Most of the time such activities are in accordance with and for the benefit of businesses, such as hotels and restaurants. The big difference, however, is that while vigilantes are more spontaneous local short-term groupings, paramilitary organizations have a relatively permanent nature, span a larger area or even the entire country, are highly coordinated, and their activities include much more than simply private 'justice making', crime control or social cleansing.

Warlords, according to Robinson (2001), can be defined according to the following criteria:

1) operate in a collapsed or significantly collapsing state;
2) pursue a narrow, commercial self-interest;
3) have access to armed force;
4) resort to violence to protect self-interest, regardless of the needs of the society in which they operate and contrary to the fundamentals of international human rights law; and
5) operate outside any democratic mandate or accountability.

Unlike the gangster or criminal, who routinely employs small arms, the warlord has access to and employs large quantity of weapons and has some sort of land and air mobility. And, unlike the insurgent, the warlord operates without popular support and threatens the population indiscriminately if it suits his aims (Robinson 2001).

Paramilitaries resemble warlords in the centrality of the economic motivations behind their acts of violence (for example, violent dispossession). However, one of the principal factors associated with the existence of warlordism, emphasized in the scholarship, is state decomposition. Most authors highlight that warlords exist where the state is extremely weak or disintegrating. By contrast, the paramilitary is a violent entrepreneur who exists alongside a strong state's coercive apparatus, thus perhaps illustrating the 'commensal relationship' (that is, one where the state and the violent entrepreneur exist in harmony to their mutual benefit), recognized by Rich (1999) and Robinson (2001). The other difference between paramilitarism and warlordism is that the latter implies a context of war and illegality. However, paramilitary groups today are increasingly present in legal economic activities in various ways, by investing in legal businesses or being hired by capitalist enterprises (local and foreign) to deal with labour unionists, populations occupying strategic territories, and other impediments to the expansions of their operations.

Warlordism is associated with the commodification of violence. Paramilitaries cannot be equated with warlords because the context within which they operate is not characterized by an absent state where the means of violence are equally available to everyone. In the Colombian context we can speak of the partial privatization of violence but not of a complete commodification because there is military equipment that is not equally available for purchase to any armed actor (for example, the guerrilla), but can only be accessed by the state and the groups it is complicit with (that is, the paramilitary).[31]

In sum, what all of these other armed actors have in common with the paramilitary is that they protect private interests (local and/or foreign), involve the outsourcing or partial privatization of violence, and blur the line between state and non-state. However, while the paramilitary shares elements with each of them, it cannot be reduced or equated with any one of these actors alone. The special feature that distinguishes it is its pervasive penetration of the state – in other words the presence of paramilitary power inside major state institutions (beyond the coercive apparatus) which in turn shapes the state's decision-making, policies and practices.

Conceptualizations of the Phenomenon of Paramilitarism in Colombia

In order to analyze the ways in which the phenomenon of paramilitarism in Colombia has been conceptualized, it is first necessary to grasp how the context in which it exists (that is, the political model in Colombia and the power dynamics of the armed conflict) has been characterized. Table 2.1 summarizes this.

The first two depictions in the Table attribute the armed conflict to the existence of a guerrilla force, without properly investigating the deeper structural features of Colombian society (e.g., poverty and sharp social inequalities), of which the guerrilla is a symptom. They ignore the state's relationship to the paramilitary and its involvement in human rights violations. The second depiction, which posits the state as a neutral third actor, is exemplified extremely well by Gaviria et al.'s (2008) work in which former President Uribe is praised for being the first to dare to combat the paramilitaries and eventually manage to make them demobilize. At the same time he was brave enough to stand up to the FARC and did not hold negotiations because, according to Gaviria et al., he would not allow criminals to dictate the conditions for negotiation. Such perspectives on the conflict are clearly not useful for understanding paramilitarism. The third depiction recognizes the state's human rights abuses and its links to

Table 2.1   Depictions of the Colombian Armed Conflict

View	Description
The Colombian State versus the Guerrilla	The state wages a war on the predatory terrorist narco-guerrilla.
Left-wing Guerrilla versus Right-wing Paramilitary	The conflict is between left-wing and right-wing illegal armed groups that engage in drug-trafficking; the state condemns and persecutes both equally in order to put an end to the violence and to protect the civilian population.
The Colombian State as a Human Rights Violator	The state and the paramilitary are complicit in human rights violations. Both use the existence of the guerrilla as a pretext to repress social movements, activists, advocates of social change and human rights, and all who oppose the current politico-economic model.
Low-intensity Democracy	The state is officially democratic but coexists with forms of parainstitutional violence; it is led by a modernizing transnational elite which promotes the state's indirect role in repression by tolerating paramilitary forces.
The Parainstitutional State	The state is characterized by a repressive military apparatus and a political system that is penetrated by powerful capitalist factions and their paramilitary groups.

paramilitary forces, however it does not rest on a solid historical analysis of the political and economic processes that have paved the way for the emergence of paramilitarism. Thus, while such a presentation reveals ongoing violations by armed actors against civilians, it does not invite us to view the conflict in relation to the development of capitalism in Colombia. The last two portrayals of the conflict are useful in the sense that they hint at the duality of the Colombian political model – formal democracy underlain by parainstitutional forms of violence. This is in turn very helpful for understanding the relationship between the paramilitary and the state, particularly the penetration of the state by paramilitary power as well as the use of state power to facilitate and at times give legitimacy to parainstitutional violent activities (as will be shown in Chapters 3 and 4).

The following pages review existing conceptualizations of Colombian paramilitarism[32] that include explanations of how and why paramilitary forces emerged. Several types of characterizations of the paramilitary can be found in the theoretically oriented literature:

1) the paramilitary as a logical outcome of a weak state;
2) the paramilitary as a criminal actor that is a product of a weak state; and
3) the paramilitary as the unofficial instrument of an allegedly democratic state which nonetheless relies on violence to deal with dissent.

Some variations can be found within each of these approaches. One of the most common factors to which the emergence of paramilitarism in Colombia has been attributed by Colombian, Latin American and a few Western scholars has been the weakness of the state – understood as the state's lack of monopoly over legitimate force for the sake of collective security and its inability to consolidate democracy by establishing the rule of law. Some authors, such as Bejarano and Pizarro (2002), have gone as far as to argue that in the 1990s the Colombian state suffered a partial collapse where the political system could be described as a 'besieged democracy'. Within the 'weak state' thesis, two models have been put forward with regard to how and why paramilitary forces emerged under the conditions of a weak state. The first argues that the state itself outsources and partially privatizes access to the means of violence. Kalyvas and Arjona (2005) claim that the formation of paramilitary bodies is directly related to the construction of the state. Strong states do not need to privatize or outsource violence since they can control or repress threats in an effective way using their police apparatus. Weak states, on the other hand, cannot do the same. In Kalyvas and Arjona's view, states today must dissolve their monopoly on violence in order to preserve it. These authors offer a typology to classify the formation of paramilitary organizations based on an intersection of state resources (high/low) and the kind of threat (big/small). Under circumstances of high threat and high resources, paramilitary armies are formed outside the formal structures of the state's apparatus because the latter is incapable of facing the irregular nature of the threat from the guerrilla and thus has to resort to outsourcing.

The other version of the 'weak state' perspective attributes the emergence of paramilitary forces to the initiatives of private individuals who feel that the state is unable to guarantee their safety and security. According to Tobon (2005), the existence of paramilitary groups is a clear indication of serious faults in the relationship between state and society. He sees the paramilitary as self-defence associations which were created by a particular social sector as the only way to ensure its survival, and which eventually acquire a political and territorial dimension. Romero's

(2000; 2003) extensive writings on paramilitarism rest on a similar logic. His work is an example of a study that focuses on a particular time period (1980s–90s) and a specific geographic region (Department of Córdoba), yet claims to offer an explanation of the phenomenon of paramilitarism as a whole. Romero highlights specific state policies and events which he contends were perceived by sectors of the elite as a political opening that favoured the guerrilla as well as its allies and sympathizers. These were: the peace talks with the guerrilla initiated in 1982 by President Betancur; the first popular election of mayors in 1988; the new Constitution of 1991; and the peace talks with the FARC initiated in 1998 by President Pastrana. Romero (2000) explains that in organizing to defend and protect themselves from the guerrillas and common criminals and to oppose the reformist policies of the central state, these elites developed strong social ties and a shared vision of a corporatist social order and their place within it. 'This network of camaraderie and solidarity shaped a political identity that resisted state penetration, collective mobilization, and autonomous peoples organizations – promoting masculine values of courage and honour, and relying on retaliation to resolve conflict' (Romero 2000: 52). Romero argues that the consolidation of paramilitary forces debilitated the authority of the central government in areas where they exercised dominance and consequently exacerbated the decline of the Colombian state. Romero's more recent work (2007) seems to invert this argument slightly by claiming that the territorial expansion of the AUC (where the military and police failed to impede or confront AUC operations) was made possible due to the functional collapse of the state and its political representation. At best, one may conclude here that possibly the author sees the destabilization of the state and the expansion of the paramilitary as two simultaneous processes mutually reinforcing each other. In any case, what is certain is that Romero perceives a definite relationship between the weakness of the state and the existence of paramilitary groups.

Cubides (2005) also insists that the existence of the paramilitary can be explained as a response to the threat from the guerrilla. In the same vein, Rangel (2005) sees the predatory practices of the guerrilla in the mid 1990s, combined with state impotence, as the impulse for the formation of paramilitary bodies. A more simplistic proposition, yet one which essentially echoes the same argument that the paramilitary emerged due to the state's inability to protect its citizens from the guerrilla, is offered by Spencer:

The core of their [the paramilitary's] intense violence is the pent-up anger and frustration of important sectors of the rural population at guerrillas who have terrorized the countryside for 30 plus years. This has been exacerbated by a state that has been unable to provide more than fleeting relief from insurgent violence. The continual inability of the government to bring peace or provide adequate protection to the population in the rural areas has provoked people to take matters into their own hands and protect themselves against the insurgents. (2001: 3)

Moreover, Spencer states that, while the guerrilla is an economic predator who tears down public authority, the paramilitaries restore law and order as well as economic stability. Also aligned with the view that the paramilitary has emerged from within civil society in response to the threat from the guerrilla is Gaviria et al.'s definition of paramilitarism as the 'illegitimate use of structures and discipline of military style to combat criminal organizations such as the guerrilla' (2008: 91).

Another way in which paramilitary groups have been characterized has been as criminal organizations. This approach is not in disagreement with the notion of the 'weak state' and, in fact, some of the authors believe that the state's weakness and subsequent need to outsource has enabled the formation of criminal networks. Duncan (2006) sees the paramilitary or *auto-defensas* as warlords or criminals who seek to impose a mafia regime over the state. In McLean's words, 'Simply put, the guerrillas and their enemies, the paramilitaries, are all outlaws' (2002: 130):

The Colombian state is not providing the security its citizens have a right to expect, which is a fundamental failure. The rate of kidnappings remains intolerably high. Guerrillas regularly bomb an oil pipeline responsible for a significant portion of the country's export revenues; seem more intent than ever on destroying highways, bridges, and electrical grids; and are now threatening reservoirs and other crucial infrastructure. Paramilitary groups have grown ... to fill the vacuum left by the absence of government. Following the guerrillas' example, they also began making money from criminal enterprises. (2002: 132)

While the next view of paramilitarism sees it primarily in terms of its criminal dimension, it nevertheless recognizes its potential political implication. Zuleta (2005) argues that paramilitary groups are private armies of assassins that constitute the drug-trafficker's clandestine apparatus of repression. The gradual growth of this private apparatus of repression and its eventual incorporation into the fabric of society through the recruitment of poor youth (who cannot achieve access to goods and services within conditions of legality) into its networks has led to the devalorization of life and 'the conversion of death into a regular source of income for some' (Zuleta 2005: 126).

The work of Medina (1990) and Medina and Tellez (1994) offers a perspective that incorporates elements of the argument that characterizes the paramilitary as a criminal formation that is a symptom of a weak state. However, these scholars attempt to use greater precision when characterizing the paramilitary by avoiding the tendency to lump such armed groups in the category of organized crime. Instead, their work stresses what they believe to be the usurpation of the paramilitary phenomenon (founded upon national security and anti-communism discourses) by drug-trafficking. In turn, this has 'generated a phenomenon with great destabilizing capacity socially and politically – narco-paramilitarism' (1994: 49). Medina and Tellez explain that this 'detouring' of the initial objectives of paramilitarism in the 1980s is part of a historical development of parainstitutional practices since the 1950s, prompted by the incapacity of the state to resolve old and new problems and to accept the challenges of social transformation which in turn have led it to the confront conflicts with force (that is, the institutionalization of violence). The authors conclude that

> The cause that generated parainstitutional practices in the 1950s is the same as those that reactivated these practices in the 1980s: the incapacity of the state to maintain the public order as a consequence of a profound crisis of governability. The difference is that bi-party violence was displaced by multiple violences which today together form a much more complex phenomenon. (1994: 79)

Palacio[33] is the scholar who has most accurately touched upon the various emerging dimensions and tendencies of paramilitarism, even at the time he was writing 20 years ago. He explains that while the existence of narco-paramilitary groups signifies a breakdown of the state's monopoly over legitimate physical force, the 'state is forced to recognize the reality of parainstitutionality and to enter into negotiations with parainstitutional actors ... Such negotiations take the state beyond the legal limits of state

actions. In other words, the state itself engages in parainstitutionality'
(1991: 106–7).

The next category of academic literature, where we can find characterizations of the Colombian paramilitary, consists of a number of progressive
works mostly published in the English-speaking world, such as Giraldo
(1996), Dudley (2004), Stokes (2005), Toledo et al. (2004), Murillo and
Avirama (2003), Livingstone (2004) and Mazzei (2009). What all these
works have in common, besides revealing quite well the seriousness of
human rights violations by right-wing armed groups, is that they portray
the latter as the privatized right-hand of the state created to do its 'dirty
work' (subcontractors of state terror) in the face of international concerns
over human rights. Giraldo (1996) expresses well this idea by stating that
linking 'the civilian population to armed actions ... hides the identity
of state agents and/or allows them to carry out covered up operations'.
Consequently, human rights violations 'cannot be attributed to persons
on behalf of the state because they have been delegated, passed along or
projected upon confused bodies of armed civilians' (Giraldo 1996: 81).
Similarly, Stokes (2005) states that the deployment of paramilitary forces
has allowed for a distancing between official state policy and the unofficial
use of terrorism directed against the civilian population.

Mazzei (2009) advances a theory of paramilitary emergence based on her
studies of El Salvador, Mexico and Colombia. Within this third approach,
her work offers by far the most elaborate way to conceptualize paramilitary
groups. Mazzei recognizes that in the countries she studied there is a
mutually beneficial relationship between the political elite, the economic
elite and the military. The latter is financially sustained by the economic
and political elites and serves to protect their interests by dealing with
uprisings and dissent. However, eventually, overt repression (commonly
relied upon in the past) becomes a less viable option due to the global
pressures around human rights issues. In the context of growing social
mobilizations demanding social change on the one hand, and the need for
international support on the other, a split between political, economic and
military circles takes place, where hard-liners find themselves 'in search of
extra-institutional means of achieving their preferred ends' (Mazzei 2009:
17–18). Mazzei's 'triad' model illustrates the confluence and conditions
under which paramilitary organizing is facilitated. It is composed of
factions of the economic elite, who provide finances, training sites, and
other organizational necessities; factions of the political elite, who provide
political and legal 'cover', ideology, purpose and leadership; and factions of

the military or security forces, who provide arms, training and leadership. A crucial point here that Mazzei makes is that

> The potential organization and success of paramilitaries is in part determined by the repressive history and capacity of the state; the historical willingness of the state to use repression is likely to provide paramilitaries with a sense of legitimacy in continuing tactics that have been successfully used in the past. This underlies the sense of the right to use repression (2009: 18).

Table 2.2   Conceptualizations of the Paramilitary in Colombia

Approach	Views	
The Paramilitary as the Outcome of a Weak State	i)	The state outsources/partially privatizes violence in order to face the threat of the guerrilla.
	ii)	Certain social groups within the elite create self-defence associations (private armies) to protect their security and interests.
The Paramilitary as a Criminal Actor	i)	Paramilitary groups constitute an illegal apparatus of repression that is part of the illegal business of drug-trafficking.
	ii)	Paramilitarism in Colombia today is 'Narco-paramilitarism' which is the result of traditional paramilitarism (based on a counter-insurgency ideology) being usurped by drug-trafficking.
	iii)	The Parainstitutional State – narco-paramilitary forces exercise power over the state and consequently the state itself engages in parainstitutionality by having relationships with these groups.
The Paramilitary as Subcontractors of State-terror	i)	Paramilitary groups enable the state to remain democratic in theory while being complicit in gross human rights violations; the economically powerful sectors support and participate in these groups.
	ii)	The 'Triad' Model – factions of the economic elite, the political elite, and the military make possible the formation of paramilitary forces.

Critique of the Existing Conceptualizations of Colombian Paramilitarism

Each of the characterizations that I discuss below contains some interesting insights, but on its own is incomplete. Some insights need to be integrated into a broader framework in order to illuminate the entirety of this complex multi-dimensional phenomenon.

The 'Weak State' Approach: Rendering Invisible the State-Paramilitary Alliance

The first approach discussed in Table 2.2 is based on the notion of the 'weak state' (employed by Romero 2000, 2003; Spencer 2001; Cubides 2005; Tobon 2005, Rangel 2005; and Richani 2007), where the paramilitary is seen as an outcome of either the outsourcing of violence by the state or the initiative of sectors of civil society who feel that the state is unable to guarantee their safety and security and consequently form private armed groups. The weakness of the Colombian state has been generally described as its inability since the nineteenth century to command the totality of its national territory. It is believed that this factor, combined with the development and extension of guerrilla forces in the twentieth century, led to the subordination of the state and its armed forces to private armed groups. There are several important problems with this approach.

To begin with, the term 'weak' has not been defined in a rigorous manner by many of the scholars of paramilitarism who rely on it. On the other hand, state capacity (which is what the 'weak/strong' categorization has to do with) is something that has been theorized at length and in a variety of ways.[34] Tilly (2003) uses the term 'government capacity' to refer to the extent to which government agents control resources, activities and populations within the government's territory. Combinations of different degrees of government capacity and democracy produce different types of regimes, which in turn affects the character and intensity of collective violence within them.[35] Tilly (2005) argues that, in general, levels of collective violence run higher in low-capacity regimes, whether democratic or non-democratic, the reason being that high-capacity governments generally limit independent access to coercive force and can prevent the acquisition of lethal arms by any group.[36] According to him, Colombia can be classified as a high-capacity regime containing significant zones that escape the control of the central government.

Scholars of Colombian paramilitarism who rely on the 'weak state' notion have not explicitly drawn on any specific dimensions of state capacity outlined by the different theories on this subject. Most seem to hint at the weakness of the state in performing its coercive functions by pointing to military spending or the lack of armed presence in all parts of the country. Campbell also finds the idea of a 'weak state' problematic. He argues that

> all too often it amounts to the expression of an ideological bias: 'weak' states are those that are simply not as 'democratic' as 'we' are. It also implies a strict dichotomy between 'weak' and 'strong' states that usually does not reflect reality. Modern states, even 'weak' ones, are complex and given the multitude of functions even relatively feeble states have to fulfill today, it is quite possible for a state to be strong in some areas and weak in others. (2000: 12)

Apart from the lack of conceptual clarity around the notion of a 'weak state' when used by Colombianist scholars, if we accept the term 'weak' in reference to the state's coercive capacity,[37] then we find a second and more substantial problem in the formulation that paramilitary groups emerge in response to the weakness of the state. As Chapter 4 will demonstrate, throughout the twentieth century the state's coercive apparatus in Colombia began to expand and enhance its functioning, especially since it became the second largest recipient of US military aid, enjoying consistent support for training, new military technology, and bases. It would no longer be accurate to label it as weak, at least as far as the last 20 years are concerned. In 1985, the military numbered 185,000. By 1996, it was 266,000. By 2010, the military had become the largest employer in the country with 450,000 members, representing 50 per cent of all government employees (Richani 2010). The Defence and Democratic Security programme is an example of how the state's coercive apparatus has been extending its reach beyond its institutions and incorporating members of civil society into its networks. Furthermore, between 2005 and 2006 there was a 54 per cent increase in government-initiated battles against the FARC and the government has been successful in establishing some form of security presence in areas where it previously had none (Holmes, Gutierrez and Curtin 2008). Thanks to new monitoring and radar interception technology provided by the US, the Colombian state was able to detect the precise location of the FARC camp in Ecuador, and to subsequently launch an aerial attack on 1 March 2008, followed by an assault with ground troops in which FARC deputy commander Luis Edgar Devia Silva (alias Raúl Reyes) was killed along with other guerrilla members and four Mexican students. A similar example was the operation known as Sodoma, in which 20 planes and 37 helicopters were employed to discharge 50 bombs on a FARC camp near La Macarena, Department of Meta, on 23 September 2010, killing the FARC commander Jorge Briceño Suárez, alias Mono Jojoy (El Colombiano 2010).

Another point of critique that can be directed at the 'weak state' approach concerns its characterization of paramilitaries as self-defence forces created by certain sectors of civil society in the 1980s in response to the failure of the state to ensure security and protection for its citizens (a position common to Rangel 2005, Romero 2003, Pizzaro 2004, and Tobon 2005). Romero (2003) goes so far as to argue that the consolidation of paramilitary forces has led to the decline of the Colombian state by debilitating its authority. These accounts appear to be completely oblivious to the creation of paramilitary units by the state itself in the 1960s, which played the crucial role of laying down the legal and military foundation for the existence of paramilitarism. Moreover, they render invisible the subsequent long-lasting and mutually beneficial relationship between the Colombian state's coercive apparatus and the paramilitary as well as the presence of paramilitary power inside the state.

The above explanation also does not question the appropriation by capitalist classes of the term 'self-defence', which was originally used by poor peasants in the 1950s and 1960s as they organized to protect their lands against the incursions of *hacienda* owners. Authors who unquestioningly accept the paramilitary's use of the label 'self-defence' tend to treat the creation of paramilitary units in the 1980s as being solely a matter of self-protection. However, far from merely offering protection, the objective of these groups has been to actively seek out and exterminate any potential guerrilla sympathizers (whether real or imagined), including labour unionists, peasant leaders, progressive students, educators and politicians, as well as to destroy their organizations. Terror has been used 'not only to debilitate the enemy but also to break, prevent and impede the links between the population and the enemy (repressive or dissuasive terror)' (Lair 2003: 96). Pearce also eloquently challenges the 'weak state' perspective, pointing to the positive relationship between the state and illegal armed groups:

> While violence in Latin America is often treated as 'state failure,' we may in fact be seeing something more dangerous, the emergence of particular forms of the state, dedicated to the preservation of elite rule, at times combating and at times conceding space to aggressive new elites emerging from illegal accumulation, in which permanent

violent engagement with violent 'others' plays into the broad project. (2010: 288)

In sum, the decentralization of violence and the state's lack of monopoly over the means of violence cannot be equated with state weakness. In the case of Colombia, the state outsourced violence by arming civilians for the interest of certain sectors of society, and later on allowed this sector (that is, the economically dominant classes) to create their own armed groups. This is not illogical, since the state is allied with the wealthy classes and the latter in turn have access to political power. After all, as Miliband (1973) powerfully argues, state institutions function to protect and serve capitalist interests. The state's role is to sustain the current economic order. In an order characterized by class inequality, the state can never be neutral since it prioritizes the interests of one class over another. Right-wing armed groups continue to operate not because the state is willing but unable to eliminate them, rather, the state allows them to exist because this arrangement contributes towards preserving the capitalist system. Hence, paramilitarism is a strategy of the state-capital alliance, rather than the unintended outcome of a state that is incapable of limiting the access of private groups to armed force.

The Paramilitary as a Criminal Actor: Obscuring the Political Motivations behind Paramilitary Violence

The conceptualization of the paramilitary as a criminal actor also relies partially on the 'weak state' notion when it argues that such groups constitute an illegal network or mafia that emerges in the context of a failed state (for example, Medina and Tellez 1994; McLean 2002; Duncan 2005; Zuleta 2005). The difference with the previous approach is that paramilitary organizations are seen not as legitimate self-defence bodies but as criminal actors. More specifically, they are viewed as an illegal agent of violence that constitutes an inherent element of the drug-trafficking business. While Medina and Tellez (1994) recognize the existence of 'traditional paramilitarism' prior to the 1980s, they argue that since then paramilitarism has been subverted by a shift from its original objectives (i.e., defeating the guerrilla) towards ensuring successful drug-trafficking operations by solving problems (such as settling accounts) that formal regulated institutions could or would not address.

The fundamental flaw of this argument is twofold. Firstly, with its emphasis on the paramilitary's involvement in drug-trafficking, it ignores the attacks of paramilitary groups in the 1980s and 1990s on social movements and popular organizations' leaders and members. During that time, paramilitarism was not external to the functioning of the state but rather an integral element (albeit unofficial) of its national security strategy. The counter-insurgency ideology that has guided the paramilitary's selection of its targets has not disappeared just because drug-trafficking as an illegal economic activity has grown in importance.[38] Paramilitary groups and the state's coercive apparatus have continued to operate in mutually supporting ways, as Chapters 3 and 4 reveal.

Secondly, and even more importantly, Medina and Tellez's claim fails to recognize that drug-traffickers are capitalists who, in addition to their illegal activity, also invest in properties and enterprises. Unionists, human rights activists, peasant leaders and anyone who seeks social justice interfere with their interests and are considered potential enemies. Thus, drug-traffickers do not differ from other capitalist classes with regard to their economic interests and their need to neutralize those who represent obstacles or challenges to processes of capital accumulation (for instance peasant movements for land recovery or unionists struggling to reduce the rate of exploitation). The fact that drug-traffickers resort to violence to punish traitors or competitors does not negate their ability to also resort to violence to advance their class interests against anyone who constitutes a threat. As with any other capitalist group, drug-traffickers have political interests that need to be defended. (A more elaborate discussion of the intersection of paramilitarism and drug-trafficking will be undertaken in Chapter 5.)

The characterization of the paramilitary as an apolitical criminal actor collapses the notion almost completely into the category of the warlord. As discussed earlier, warlords operate within markets of violence (that is, economic fields that entwine violent and non-violent forms of appropriation and exchange). The notion of markets of violence is very useful in that it draws attention to the economic motives behind the activities of the paramilitary. However, it also presumes a disintegration of the state, which is not the case in Colombia. Perhaps the Colombian experience teaches us that markets of violence can coexist with a strong state's coercive apparatus. The other limitation of the concepts of warlords and markets of violence is that they create an impression that violence conducted for the generation of revenue only takes place within the realm of illegal economic activity. This is also the implication of Medina and Tellez's (1994) argument that paramilitarism is confined to drug-trafficking. But paramilitary violence has been unofficially employed by many legal enterprises, including foreign-based companies operating in Colombia such as Coca-Cola, Drummond, and Chiquita. In sum, in this approach (which views the paramilitary as a criminal actor), drug-traffickers are not only treated as apolitical but are also presumed to operate outside the capitalist economy, which is of course not true.

The other argument within this approach is put forward by Palacio (1991), who views the paramilitary as a criminal actor but also recognizes its positive relationship to the state. Because the state engages with this parainstitutional actor, Palacio calls it a 'parainstitutional state'. This perspective offers a valuable insight by hinting at the back-and-forth mutually beneficial exchanges that take place between paramilitary groups and members of state institutions. However, one of the limitations of Palacio's work (due to the period in which he was writing) is that it only focuses on cocaine entrepreneurs as the basis of this 'parainstitutional state', while over the last 15 years, agribusinesses and extractive industries have become an increasingly important economic base for the elite employing paramilitary forces.

The other shortcoming of Palacio's concept of parainstitutional violence is that it makes it possible to lump together paramilitaries and guerrillas in the same category of parainstitutional actors. The implications of equating the guerrilla and the paramilitary, in addition to erasing some fundamental structural and historical features of Colombian society, also conveniently serves to situate the state as a neutral third actor combating both forces equally. In fact, Sánchez explicitly expresses this belief that 'Here we have in a nutshell the nature of the Colombian crisis: two opposing rivals, against an absent enemy, the State' (2001: 25). Two crucial factors seem to be disregarded here. One is the control the paramilitary elite have over the best land, drug-trafficking, many criminal organizations, and numerous legal businesses. The guerrillas, in contrast, are not large-scale landowners, cattle-ranchers, owners of lottery establishments, shopping centres, illegal drug laboratories and trafficking routes. The second is the continuous historical relationship that has existed between the paramilitary and the state. For instance, not a single unit in the police or military has been formally or specifically dedicated to fighting the paramilitary. While both the guerrilla and the paramilitary are outside the state, the former fights against the state while the latter targets those who are against the state. It is difficult to deny that the guerrilla's relationship to the state and economic resources bears no resemblance to that of the paramilitary.

The Paramilitary as a Subcontractor of State Terror: Human Rights Issues Overshadow Class Inequality

The last approach to conceptualizing the paramilitary, reviewed earlier, emphasizes its repressive role as an instrument of the state. The majority of authors such as Murillo and Avirama (2003), Dudley (2004), Toledo et al. (2004), Livingstone (2004), Stokes (2005), and Aviles (2006) do an excellent job of revealing the magnitude of human rights violations carried out by the paramilitary and the state. Some of them recognize the use of violence by such groups to advance private economic interests. Others, such as Aviles, highlight the privatization of repression as a tactic used by the state to make itself look good in the eyes of the international community. They do not, however, discuss the role of the paramilitary in capitalist activity through dispossession and the repression of social movements. In general, works in this third category fail to offer a sufficient analysis of paramilitarism as a phenomenon on its own, preferring simply to mention the paramilitary as merely one of the actors in the armed conflict. Of course, this is not a weakness in so far as the goal of these works is to throw light on the armed conflict as a whole, and especially the role of the US therein, rather than on the paramilitary specifically or in any depth.

The other issue here is that while these accounts are eye-opening with regard to human rights abuses, what remains under-investigated is the extent to which the paramilitary currently exercises domination over state institutions by offering state employees lucrative opportunities in exchange for their collaboration. Portraying the paramilitary mainly as an instrument at the disposal of the state and dismissing the fact that today right-wing armed groups provide extra jobs for state employees can obscure the paramilitarization of the Colombian state and make it difficult to grasp fully the ways in which the state's coercive apparatus and the paramilitary complement each other.

Writings such as those by Sánchez (2001) reveal human rights violations carried out by the paramilitary and/or state forces but lack any class-based analysis and thus miss the centrality of paramilitary violence to processes of capital accumulation. In this approach, the term 'class' is employed only occasionally, usually when referring to the objectives of the guerrilla, and the notions of capital accumulation and class struggle are regarded as remnants of an old-fashioned Marxism that have no place in today's world. Relying solely on a human rights framework, however, does not provide us with any means of identifying the ways in which the profoundly uneven class structure fuels violence. Thus, although such works condemn the abuses against civilians, they do nothing to expose the systemic and structural causes of paramilitary and/or state-sanctioned violence which is in essence aimed at reproducing capitalist social relations. Consequently, paramilitarism has been detached from the capitalist social relations that it serves to sustain. However, even though the composition and dynamics of the different social classes in Colombia have changed over time, the realities of class inequality and class struggle remain, hence the need to examine paramilitary and state violence in this context. Those scholars who rely solely on a human rights framework without a class analysis tend to focus primarily on violations of the rights to life, liberty, physical integrity, security of person, and freedom of thought and conscience. I believe, however, that it is important to take into consideration all spheres of human rights, not only the political and civil ones, but also the economic and social ones, as mentioned in Article 25 of the UN Universal Declaration of Human Rights:[39]

> Everyone has the right to a standard of living adequate for the health and well-being of himself and of his family including food, clothing, housing and medical care and necessary social services, and the right to security in the event of unemployment, sickness, disability, widowhood, old age or other lack of livelihood in circumstances beyond his control. (UN 2012)

This is why I argue that a Marxist political economy framework, with its attention to the material foundations of social relations, can help us in understanding the causes of human rights violations in a more comprehensive way.

Within the literature on the paramilitary as a subcontractor of state terror, Mazzei's (2009) 'triad' model gives the most consideration to the importance of the local elite in the formation of paramilitary groups. Because her work is of a comparative nature (looking at the commonalities of the paramilitary experience in Mexico, El Salvador and Colombia), it does not venture into exploring how these groups penetrate, influence and control the state, or into examining the transnational dimensions of the paramilitary's involvement with capital (including with foreign companies). Mazzei uses the term 'factions' when referring to the involvement of the political elite, the economic elite and the military. This serves the purpose of providing a balanced perspective that recognizes that not everyone in a given social sector engages in paramilitarism. Yet the Colombian experience shows that these factions are not steady, but growing. Once paramilitary services exist, larger and larger parts of the wealthy sector end up implicated with them either by hiring them directly or by being forced to pay protection fees.

3

From Colonialism to Neoliberalism: A History of Dispossession

Two uninterrupted, intertwining themes run throughout Colombia's history: 1) social relations marked by inequality, exploitation and exclusion, and 2) violence employed by those with economic and political power against the working majority and the poor for the purpose of maintaining control over resources and labour, and eliminating or suppressing dissent. As Oquist (1980) remarks, violence has been an important and often decisive social process in the structuring of Colombian society. This chapter traces the major patterns of class formation, conflict, and the evolution of the state from the 1500s to the present by looking at four historical periods: 1500–1810; 1811–1902; 1903–89; 1990–2014. Each period represents the emergence of new social forces and an increase in the importance of violence and repression. The chapter then examines the paramilitary-state alliance prior to the demobilization (1990–2006) and ends by commenting briefly on the condition of social movements today.

Planting the Seeds of Poverty and Inequality: Conquest and Colonialism (1500–1810)

The Creation of Propertylessness: Slaves, Peones and Terrajeros

The Spanish conquest of the territories that form contemporary Colombia began in the early 1500s.[1] During the first century of the Spanish colony of Nueva Granada (today's Colombia), the colonizers' primary objective was to accumulate wealth by exploiting the land. This took the form of two main activities that became the basis of the colonial economy – the operation of gold mines and the establishment of *haciendas*[2] through land concessions that were given by the Spanish Crown to individual landholders (Pearce 1990). By the 1700s, large landholdings were used for agricultural produce, raising cattle and sugar plantations. Other economic actors, besides landowners and mine owners, included the owners of riverboats (the economy depended on water transportation at that time) and merchants (Oquist 1980).

The productive relations in Nueva Granada from the 1500s to 1700s were determined by three factors: 1) abundance of minerals, gold and land; 2) heavy reliance on human labour; and 3) a critically short supply of labour. At first the institution of the *encomienda*[3] was used to force indigenous people to work in the mines. Between 1503 and 1660, 185,000kg of gold was extracted by forced indigenous labour (O'Connor and Bohorquez 2010). This, however, was a temporary solution to the labour shortage and eventually contributed to the demographic disaster among indigenous people caused by wars, European diseases and the reduced birth-rate due to the disintegration of their communities. Among the indigenous most quickly decimated were those employed on the mines and riverboats, and those carrying cargos from ports inland (Safford and Palacios 2002). The shortage of labour was addressed in two ways. The Spanish Crown abolished indigenous slavery and established the *resguardo*[4] in 1592. However, the Spanish in Nueva Granada felt that the *resguardos* allowed the indigenous to subsist independently and reduced the availability of labour for colonial agriculture and mines. Thus, the Spanish-Americans began to administer *resguardos* independently from the Spanish Crown by lumping the population from several *resguardos* onto one so that there would not be enough land for all indigenous residents to engage in subsistence agriculture. This forced some to become *terrajeros*[5] on *haciendas*. The other solution to the labour shortage was the introduction of African slaves into the colony (Oquist 1980).

The productive relations in Nueva Granada from the sixteenth century to the eighteenth were characterized by the emergence and consolidation of patterns of highly uneven land distribution (visible in the *latifundio-minifundio*[6] structure that has continued until today) as well as patterns of interrelated class and racial inequalities. The most privileged class was the *criollos* (individuals of Spanish descent born in Nueva Granada who were landowners, merchant capitalists and high clergy). The artisans and peasants on *minifundios* were usually poor individuals of Spanish descent, mulattoes (those of mixed African and European ancestry), and *mestizos* (those of mixed indigenous and European ancestry). *Hacienda* labourers (including *peones* and sharecroppers) were commonly *mestizos*, and domestic servants were mostly indigenous women. After the late 1600s, the workforce on the mines, riverboats, ports and plantations consisted mostly of African slaves (Oquist 1980). Some of the indigenous population became *peones*, while others remained on *resguardos*. Those *resguardos* located in more remote areas such as the Tierradentro region, in what is today the Department of Cauca, were able to retain their land and culture until the mid twentieth century (Ortiz 1973).

Primitive Accumulation Phase I: Appropriation of Original Indigenous Land and Resguardos

From its inception, the colony of Nueva Granada was ridden with violent conflicts. Initially, the conquest of indigenous territories, the domination of indigenous people, and the imposition of Spanish authority through political and economic institutions, were achieved by violence. Wars of conquest were carried out against the more highly organized indigenous nations such as the Chibchas, while wars of dislocation and extermination were waged against the less organized and less economically developed groups due to the fact that they were more difficult to defeat militarily and to control politically, and consequently were considered less attractive from the point of view of the colonizers (Oquist 1980). Subsequently, most of the conflicts that led to violence were inter-class in nature – between the colonizers on one hand and the indigenous, African and poor Whites on the other. One of the most persistent kinds of violence during the colonial period was employed by the Spanish miners, landowners and merchants for the purpose of maintaining the institution of slavery. There were many collective slave rebellions. A movement of escaped slaves was formed in Cartagena in 1600. They raided ranches, destroyed commerce throughout the Caribbean coastal region, and formed a permanent fortified settlement known as Palenque de San Basilio. In 1619, there was a second slave revolt. Many other *palenques*[7] were formed throughout the colony and by the second half of the 1800s the conflict between slave-owners and the anti-slavery movement of slaves resembled a civil war.

The other major kind of violence employed by the colonizers was for the purpose of primitive accumulation[8] and was manifested in two processes which occurred simultaneously from the 1600s onwards. One was the displacement of the indigenous population from territories with favourable characteristics – mineral deposits, fertile land, access to water, transportation routes and so on. The separation of the indigenous from their means of subsistence allowed for the transformation of what used to be the native inhabitants' communal lands into *haciendas* and created a supply of labour. The other process of primitive accumulation was the expropriation of *resguardo* land through individual and collective attacks. It was common for indigenous families to be left with less than a hectare of land, which was insufficient to subsist on and thus they were economically coerced to work for the landlords.

A different kind of conflict had emerged between the *criollos* and the Spanish Crown due to numerous disagreements over the Spanish fiscal policy, indigenous *resguardos*, the exclusion of *criollos* from governmental posts, and Spain's role as a price-increasing yet unproductive middleman in selling to the colonies manufactured goods from more industrially advanced nations. All this discontent eventually came to a head in the Wars of Independence led by the *criollo* oligarchy against the peninsular Spanish (Oquist 1980).

Wars of Independence and the New Republic (1811–1902)

Colombia in the World Market: Export Agriculture

The Wars of Independence took place between 1810 and 1816. In 1819, the leader of the anti-colonial liberation movement Simon Bolívar made his triumphal entry into Bogotá. The Congress of Angostura proclaimed the formation of the Republic of Gran Colombia consisting of contemporary Colombia, Venezuela, Ecuador and Panama, with Bolívar as President and Francisco de Paula Santander as Vice-President. Eventually there was a separatist move by Venezuela and later Ecuador. In 1830, the Confederation of Gran Colombia dissolved. Santander succeeded Bolívar and became the President of Nueva Granada. In 1863, the current boundaries of Colombia were formed, as Nueva Granada became the Estados Unidos de Colombia[9] (Pearce 1990; Holmes, Gutierrez and Curtin 2008).

The economically dominant groups that emerged during colonialism remained in control of the wealth and political power after independence. Their vision of progress was to be achieved through modernizing their country, which implied adopting the neoclassical economic model. The latter rested on the premises that capitalist markets are smoothly functioning and self-regulatory and there is no need for the state to restrict in any way the pursuit of private bourgeois interests (Brockett 1990). By the mid 1800s, the elite strongly advocated an expansion of the export of goods associated with the primary sector – precious metals (silver and gold), tobacco, indigo, quinine and agricultural commodities (coffee and sugar). The export of tobacco and quinine created short-term booms that declined by the end of the nineteenth century (Pearce 1990). In addition to cattle-ranchers, mine owners, riverboat owners and merchants, a new propertied class emerged – the coffee bourgeoisie which included landowners, distributors and traders who made huge profits in processing and export. The nineteenth century marked the consolidation of Colombia as an exporter of goods intensive in unqualified labour and primary materials, something that continues to characterize the country's foreign trade even today. A few variations on this trend could be found in certain parts of the country, such as the Departments of Antioquia and Santander, where a large number of small-scale independent farmers (mostly poor Spanish immigrants) existed and a manufacturing industry had emerged (Pearce 1990).

Primitive Accumulation Phase II: Eviction of Terrajeros, Privatization of the Resguardos and Expulsion of Colonos

The gradual shift to export agriculture initiated several changes in the productive relations through new waves of primitive accumulation. As more and more *haciendas* began to export coffee, sugar cane and tobacco, which entailed a change in agricultural techniques, many tenants with usufruct rights to land were evicted and replaced with day labourers paid only a cash wage (Appelbaum 2003). Consequently, the expansion of landlords' estates served not only to increase the production of cash-crops but also to ensure the availability of labour by depriving peasants of their lands. Since capitalist farms could not employ all the expropriated, the problem of landlessness began to grow in significance.

Another process of primitive accumulation that unfolded during this period was the privatization of the *resguardo* – legalized by the state in 1873 through legislation that ordered each province to regulate the partitioning of its own *resguardos*, thus making possible the private capitalist ownership of these through purchase.[10] This development bears a striking resemblance to Marx's description of how in eighteenth-century England legislation was a key instrument for the conversion of 'land into a merely commercial commodity, extending the area of large-scale agricultural production, and increasing the supply of free and rightless proletarians' (1867/1990: 885). The *resguardo* law was accompanied by an ideology that claimed that indigenous possession of the land was incompatible with capitalist development and hence was an obstacle to the transition to a modern market economy (Reinhardt 1988). Once again this is highly comparable to Marx's account of primitive accumulation in England, in which he exposes the way the destruction of human life acquires legitimacy in the name of capital. 'The most shameless violation of the "sacred rights of property" and the grossest acts of violence against persons ... become fully justified ... as soon as they are necessary in order to lay the foundations of the capitalist mode of production' (Marx 1867/1990: 889).

The expansion of the large estates against the efforts of the rural poor to establish their rights to the land has been a constant theme in Colombia's history (Pearce 1990). Some historians, such as Oquist (1980), argue that the relative availability of land in the nineteenth century served to diminish economic contradictions within society. According to Oquist, peasants displaced through evictions or indigenous people who lost their *resguardos* could migrate and settle on unexploited land which was state/public domain land or land to which private individuals held title. In the latter case, the peasants had the option of becoming sharecroppers in exchange for clearing the virgin forests on the landowner's land and thus enhancing its value. Nonetheless, as Pearce (1990) points out, while there was an immense reserve of public land in the new republic, during the course of the nineteenth century, a great deal of that was granted in concessions of more than 1,000 hectares mostly to already existing landowners and a few European immigrants, but not to local *minifundistas* and landless peasants.

Many landless peasants, or those who did not want to be *peones* or *terrajeros*, tried to settle public lands (often remote or frontier territories) through slash and burn methods and engage in subsistence farming. Frequently, such squatters or *colonos* (as they were called) were expelled as part of an ongoing primitive accumulation backed up by law and coercion. Landowners who wanted to usurp a given territory on which a *colono* had settled could always find a judge ready to assert their ownership. Accompanied by the police, the landlord would order the *colono* to either leave or rent the land from him in exchange for labour services. When *colonos* invoked certain laws that were passed after 1875 to offer them some protection, the landowners ignored the laws and instead destroyed crops and bridges, formed armed groups to intimidate and terrorize the peasants, and imprisoned peasant leaders on false charges. Even though the bloody struggle over Colombia's agricultural frontier had begun (Pearce 1990), the conflict over land had not reached an acute stage yet because there were no major popular uprisings or serious challenges to the dominant classes.

During the 1811–1902 period there were eight general civil wars,[11] 14 local civil wars, many small uprisings, two international wars with Ecuador, and three coups (Pearce 1990). These wars have been characterized as intra-class conflicts that revolved around:

1)   the formation of the republican state;
2)   disagreements over the mid-century anti-colonial reforms aimed at eliminating and replacing colonial institutions by social forms of a more capitalist nature;
3)   the defence of the diverging interests of different factions of the dominant class – the traditional *latifundistas* and the Catholic Church versus the commercial and agricultural export interests.

These civil wars – prompted by conflicts of an intra-class nature in which different factions of the dominant class strived for complete control of the state – concerned more than anything questions of state organization. They did not alter the existing class relations. Even when there were partial collapses in state authority, the power of the dominant class was never threatened by the lower classes (Oquist 1980). The significant civil wars tended to break out or to be prolonged in regions that were affected greatly by a loss of demand for their products in foreign markets leading to falling commodity prices for mainly coffee, quinine and tobacco (Safford and Palacios 2002).

As the nineteenth century progressed, the civil wars increasingly took on party connotations and in turn reinforced party identification – Conservative versus Liberal. The official establishment of the parties dates back to the 1840s.[12] One of the main sources of inter-party hostility was the issue of the Church and its control over education. Conservatives were committed to the preservation of the status quo and viewed the Church as the guarantor of social order and authority. The Liberals aimed to modernize the state and critiqued the Church for being a 'bastion of privilege' (Pearce 1990: 17) which controlled education, influenced the masses, and thus helped the Conservatives gain electoral majorities. It is worth mentioning that each party also contained warring factions within itself around issues such as free trade versus protectionism and federalism versus centralism (Safford and Palacios 2002).

The penetration of the two parties into the popular consciousness, such that by the 1850s the rural and urban poor were drawn into supporting one or the other, enabled the mobilization of the population into armies of thousands of fighters for the last war of the nineteenth century – the famous War of the Thousand Days (1899–1902). The Liberals had accused the ruling Conservatives of maintaining power through fraudulent elections. The Conservatives controlled the state at that time and had a strong army. The Liberals had created numerous guerrilla armies that consisted of landless peasants, indigenous groups, and some urban poor. As Pearce explains, 'Initially, for most people participation in the local landowner's army offered promise of some favour or reward. But as they engaged in battle and saw family and friends killed or wounded by the "enemy", so loyalties and hatreds were born with deep personal roots' (1990: 20). This was the longest and most destructive of Colombia's civil wars. It resulted in the death of 100,000 (or 2 per cent of the country's population at the time) as well as great material destruction and economic chaos. It was won by the Conservative government.

According to Oquist (1980), the state organization under the newly formed independent republic in the nineteenth century was much weaker compared to its colonial predecessor. This is largely attributed to divisions within the dominant class. The early period of nation building between 1858 and 1885 has been described as an unstable federalism in which rival city-states controlled their hinterlands and clashed with one another. The state was weak and local elites relied upon the strength of local caudillos in the event of conflict. Some strengthening of the state occurred as governmental employees increased from 4,500 in the late nineteenth century to 42,700 in 1916, accompanied by the professionalization of the military and a subsequent increase in the repressive capacity of the state (Holmes, Gutierrez and Curtin 2008).

Integration Into the World Capitalist Economy and the Challenge From Below (1903–1989)

The Rise of Cattle-Ranching, Agribusinesses, Gremios and Drug-Trafficking

During the first decades of the 1900s, the Conservative government consolidated the export economy and encouraged foreign investment. US private investment in oil grew from \$30 million in 1920 to \$280 million in 1929.[13] The New York Bank provided credit to improve Colombia's import-export infrastructure, including the construction of roads, highways, railways, oleo-ducts and communication facilities. Starting in 1949, Colombia developed a close relationship with the World Bank. The latter financed the construction of the Atlantic railway (1952–61) which connected Bogotá to coastal cities, provided access to new fertile agricultural regions such as Magdalena Medio, and stimulated commercial agriculture. During the first half of the 1900s, the country also experienced a coffee boom, a growth in the number of banana plantations and cattle ranches, and the development of new extractive industries such as rubber production (Pearce 1990). In the late 1950s, the state promoted the Import Substitution Industrialization (ISI) model consisting of protectionist measures to encourage the manufacturing of goods that would otherwise be imported. Although the state did not promote industrialization as much as did its Brazilian or Mexican counterparts, it provided conditions under which entrepreneurs could take initiatives. In the 1960s and 1970s, industrial production grew and diversified in areas such as textiles, wood, paper, rubber and chemicals. The cities of Bogotá, Medellín and Cali were responsible for over 60 per cent of industrial employment (Pearce 1990).

The intensive integration of the country into the world capitalist economy through primary-sector-based exports such as coffee, sugar, bananas, meat, gold and fossil fuels (Perez-Rincon 2006) accelerated the capitalist transformation of Colombian agriculture that had begun in the nineteenth century. By the 1960s, most of the remaining traditional *haciendas* were transformed into highly mechanized capitalist agribusinesses for the cultivation of cash-crops and stockbreeding for export. Not only had many tenant farmers been evicted and replaced with wage labourers, but due to the technological advances in agribusiness operations, the demand for rural wage labourers also became lower and of a more temporary nature. As a result, an increasing number of such labourers became *colonos*, who sought to supplement their wages by engaging in subsistence agriculture. The existence of subsistence activities alongside agribusiness allowed capitalists to force occasional wage workers to bear the cost of their own subsistence and social reproduction (O'Connor and Bohorquez 2010). Many *colonos* began to settle in the Amazon region of the country in the late 1950s and eventually began to grow coca once it became a profitable crop (Chomsky 2000). Between the late 1960s and 1980, a process that Richani (2010) calls 'de-agriculturalization' was unfolding, evident in the decrease of agricultural production as a share of the GDP, the decrease in the land suitable for agricultural production, and the increase in the land dedicated to pasture (from 20.5 million hectares in 1978 to 40.1 million in 1987). By the 1980s, the agrarian political economy rested mainly on cattle-ranching and agribusiness. On the land used for agricultural production more and more was cultivated with cash-crops for export (such as coffee, plantain, banana, African palm oil, flowers and sugar cane) while less and less was used to grow crops for local food consumption such as potatoes, beans and maize (Richani 2010).

Over time, the gap between small and large producers widened and it became increasingly difficult for the former to survive. Peasant farmers lacked credit and the means to improve productivity but still produced half of the country's basic foodstuffs such as cassava, maize, plantain and potatoes. Decree 444 of 1967 gave credit and other incentives, such as machinery at subsidized prices, to large-scale producers of sugar, cotton, rice and coffee in the Departments of Tolima, Cesar and Valle del Cauca and parts of the Atlantic coast. For instance, the Agricultural Financial Fund dispensed to large farmers in one year double the amount of what the state's Integrated Rural Development Programme dispensed to small farmers in nine years. Although between 1910 and 1950 there was a growth in small and medium coffee producers in the Department of Antioquia, these were always vulnerable to intermediaries, low prices and high interest on advances. From the 1960s onwards the coffee industry was dominated by the large-scale entrepreneurs. New legislation in 1973 gave additional important concessions to landowners and financial incentives to large-scale mechanized agriculture (Pearce 1990). Parallel to these developments, the penetration of a considerable portion of the agricultural and primary resources sectors by foreign enterprises, such as the American United Fruit Company and Chiquita Bananas, took place. By 1978, 54 per cent of banana exports were controlled by two MNCs. In the 1980s mining became the main area of foreign investment (Safford and Palacios 2002). But prior to that in the 1970s a new lucrative economic activity had already appeared – the production of illegal drugs. Marijuana was gradually replaced by the even more lucrative product cocaine. The volume exported to the US grew from 15 tons in 1978 to 270 tons in 1988 (Pearce 1990).

From the early 1900s to 1946, both the capitalist class and the working class became more diversified internally. The capitalist class encompassed large coffee, sugar, rice and cotton producers and exporters, cattle-ranchers, industrialists, and bankers. The rural elites funded the electoral campaigns for politicians and the latter in turn defended their economic interests. Along the same lines, fundamental to the consolidation of the power of the capitalist class was the formation of *gremios* or economic consortiums (still in existence).[14] These were associations of landowners and entrepreneurs that guaranteed the private sector's ability to exercise influence over state policy-making by restricting the autonomy of the state and ensuring its commitment to economic liberalism. They monitored and directed state intervention in ways that favoured them and 'collaborated irrespective of party in a government dedicated to free enterprise and economic progress' (Pearce 1990: 44).

A new sector of the capitalist class emerged in the 1980s as a consequence of the illegal drug trade. Drug-traffickers invested their revenues in large swathes of arable land[15] used to raise cattle and horses in the Departments of Antioquia, Meta, Córdoba, Cauca and along the Caribbean coast (Holmes, Gutierrez and Curtin 2008). Eventually narco-capital was also invested in industry, commerce and finance, converting drug-traffickers into an economically powerful and politically influential actor (Medina 1990). Two of the leading Colombian drug-lords – Jorge Luis Ochoa and Pablo Escobar – were among the 20 richest men in the world (Pearce 1990). Another sector of the dominant class in Colombia came to be known in the late 1980s as the 'technopols' or the New Right. They were considered to be part of Colombia's TCC (to borrow Robinson's term) and were appointed to key decision-making positions within the state during the late 1980s and early 1990s. Many of them, such as Cesar Gaviria and Rafael Pardo had worked for international financial institutions or banks, defended the capitalist economy, and shared the philosophy that the capitalist free market and international competition were the keys to development (Aviles 2006).

The sectors of the rural working class included the agricultural wage labourers, *terrajeros*, *colonos* and *minifundistas*. The urban proletariat worked in the coastal and fluvial ports, on riverboats, in construction and the operation of railroads, in manufacturing, as well as in artisanal production (such as tailors, masons and shoemakers).

Latifundio-Minifundio Agrarian Conflicts

The wealth and landownership inequalities that had begun to develop under colonialism were now rapidly growing. Law 1a of 1968 resulted in a massive expulsion of tenant farmers and sharecroppers. By 1970, farms under ten hectares had diminished substantially in number and size. In the Departments of Narino, Cauca, Antioquia, Caldas, Cundinamarca, Boyacá, Santander and Norte de Santander, peasants tried to survive on diminishing plots of land alongside modern mechanized agriculture and traditional *latifundia*. In 1984 there were 1,504,215 *minifundistas* with an average of two hectares of land. Of these, 636,255 properties had less than one hectare. At the same time, ten million hectares of land were in estates of more than 500 hectares each owned by an estimated 12,000 landowners most of whom were cattle barons (Pearce 1990).

One kind of agrarian tension was between large-scale landowners expanding their properties and intermediaries/traders on the one hand, versus *colonos* and *minifundistas* on the other. *Colonos* depended on traders to sell their produce because transportation costs were too high. An agreement existed amongst the traders to buy at a certain fixed price which allowed high rates of profit for the latter. In addition, the *colono* depended on the trader for goods but had to pay on account for future harvest. Due to the high interest that traders charged, *colonos* accumulated debt rapidly. When they became bankrupt, they had to give up their land which ended in the hands of the *latifundistas* (Molano 1988). Another kind of tension was arising on behalf of indigenous *terrajeros* who gradually began to rebel against the payment of labour-rent to landowners since under the arrangement of *terraje* (rent) they could hardly find time to work on their own plots of land. A third kind of conflict was unfolding between indigenous groups that resisted the dismantling of their *resguardos* and landowners. Intense struggles also took place on large estates (especially coffee estates) where *terrajeros* or *peones* wanted to grow cash-crops on their own plots to take advantage of the opportunities (Pearce 1990).

In addition to land ownership, the concentration of wealth was visible in other areas as well. In 1985, the richest 10 per cent of urban families received a little less than 40 per cent of the total income, while 50 per cent of families received less than 20 per cent. The poorest 20 per cent gained less than 5 per cent of the total. In 1987, 0.01 per cent of shareholders from the ten biggest companies on the Bogotá stock exchange controlled 36.5 per cent of all shares. In the textile sector, two companies from Antioquia represented 44 per cent of the total capital invested. In the tobacco industry, one company (Coltabaco) controlled 77 per cent of the capital in that entire sector (Pearce 1990).

The social inequalities that began to form in the 1500s had by now matured into sharp contradictions that permeated all aspects of Colombian society and led to serious violent class conflicts in rural and urban areas. The outcome was the emergence of social movements unprecedented in size, organization and mobilization capacity. Some of these movements have continued into the twenty-first century.

Bloody Struggles Over Land: Peasant Movements, the Communist Party and State Repression

Beginning in 1910, questions revolving around access to land, authenticity of land titles, and the types and terms of rural productive relations led to the birth of organized and active peasant militancy. At first it began in parts of the Departments of Cundinamarca and Tolima where land was concentrated in the hands of a few powerful families who had owned vast expanses of land since colonial times. The Communist Party of Colombia (Partido Comunista de Colombia, or PCC),[16] became involved in peasant struggles and helped to better organize the movement which initially was about refusing to pay *terraje* to landowners and seeking an improvement in working conditions. Eventually the movement under leadership of the party developed a radical platform that called for revolutionary changes such as the expropriation of landlords without compensation. Peasants took up arms and formed their own self-defence groups in response to state repression and landlord violence in order to defend settlements of self-sufficient communities formed by landless peasants, *colonos* and *terrajeros* in parts of the Departments of Tolima and Cauca (Pearce 1990).

Another rural movement, known as the Quintin Lame (the name of its leader), consisted of indigenous poorly armed peasants. Its main objective was to eliminate the *terraje* and to protect *resguardo* lands. The movement reached its height between 1914 and 1916. It faced a strong violent response from the state and many of its leaders were jailed. Indigenous organizing (although no longer armed) continued, and in 1971 Colombia's most politically advanced organization to defend the collective and territorial rights of indigenous people was born – the Indigenous Regional Council of Cauca (Consejo Regional Indígena del Cauca, or CRIC).

A serious and effective land reform was never initiated in Colombia. There were short-term, isolated and inefficient attempts at land redistribution on a case by case basis with the objective of preserving the public order. In 1936 the first agrarian reform legislation was passed by

President Alfonso Lopez Pumarejo with Law 200. The law allowed for distribution of land that was not exploited productively. It gave ten years for this process to begin and this allowed many landowners to divide up their property among family and relatives to avoid expropriation. Liberal and Conservative landowners organized a violent opposition to the law, such as attacks on peasants who demanded their rights to land, and the law was never implemented in most areas (Pearce 1990). Later on, Law 100 of 1944 provided protection for landowners from the claims of *terrajeros* and *colonos* and facilitated violent attacks on peasant organizers in parts of Tolima and Cundinamarca by the landlords' private armed forces. In 1961, the Liberal government of President Alberto Lleras Camargo approved the Law of Agrarian Reform (Law 135) with the objective of improving the productivity of the agrarian sector and integrating it into the capitalist development of the country while at the same time pacifying existing militant peasant and guerrilla groups (Rudqvist 1983). The same law also created the Colombian Institute for Agrarian Reform (Instituto Colombiano para la Reforma Agraria, or INCORA). Mostly in the early 1970s, INCORA bought up farms totalling 472,470 hectares and distributed land to 30,000 families (Pearce 1990). Due to the sluggish and inefficient implementation of the agrarian reform, the National Association of Peasants (Asociación Nacional de Usuarios Campesinos, or ANUC[17]) was founded through the initiative of President Lleras in 1967 (Rudqvist 1983).

Under the administration of Conservative President Misael Pastrana (1970–74), laws pertaining to agrarian reform were repealed or reversed, members of the peasant movement were persecuted and imprisoned, and the dispossession of *minifundistas* and *colonos* resumed. In response to the government's harsh stand, ANUC, which had reached one million members by that time, carried out land occupations on approximately 2,000 *haciendas* between 1971 and 1975. Peasant protests and land occupations were criminalized and physically attacked by state and private armed forces. One such case took place in 1973, when 40 landless and starving families in the Department of Sucre invaded an idle terrain of 800 hectares. During a period of 11 years they were threatened, attacked, arrested, accused of subversion and taken to court. Even though at the end they were awarded some land, INCORA and the Agrarian Fund did not provide them credit or other state assistance of the kind given to private landowners and intermediaries. During the height of peasant mobilizations in the 1970s, an agrarian counter-reform led by the Conservative government was underway and was also accelerated by the investment of drug capital in landed estates. For instance, between 1983 and 1988 under the National Rehabilitation Plan INCORA bought 101,564 hectares in the country as a whole, while in the region of Magdalena Medio alone drug capital bought an estimated 180,000 hectares of fertile land (Pearce 1990).

The Offensive Against Labour Militancy

Parallel to the formation of peasant movements for land, the development of Colombia's industrial infrastructure led to a growth in the proletariat and the subsequent emergence of labour organizations and unrest. In 1918, strikes were first used as a means of labour struggle by port and railway workers in the cities of Barranquilla, Cartagena and Santa Marta, resulting in substantial wage increases. In 1919, massive artisan protests took place in Bogotá over the importation of military uniforms, leading to bloody clashes with the police. Between 1919 and 1920, there were numerous strikes by railway workers in the Department of Cundinamarca and the Caribbean coast, as well as gold miners and female textile workers in the Department of Antioquia.[18] These activities were successful in obtaining wage increases of up to 40 per cent and many other labour organizations across the country followed their example (Holmes, Gutierrez and Curtin 2008). By the end of the 1920s, the public works boom ended and many were left unemployed, which led to an increased militancy. In 1925, the National Workers Confederation (Confederación Obrera Nacional) was founded, followed by the Revolutionary Socialist Party (Partido Socialista Revolucionario) in 1926. In 1924 and 1927, there were important strikes by workers in the oil sector. In 1928, a peaceful march by banana plantation labourers in Ciénaga, Department of Magdalena, became known in the history as the Massacre of the Banana Zone (Masacre de las Bananeras). The state military violently attacked the workers and killed around 1,000 (Safford and Palacios 2002).

With the onset of the Depression in 1929, the labour movement became disempowered and weaker in its capacity to mobilize (Pearce 1990). The Liberals (seen as the progressive national bourgeoisie), under the leadership of President Alfonso Lopez Pumarejo (1934–38 and 1942–45), gradually took over the political leadership of the workers movement, as Lopez introduced many reforms in favour of urban and rural workers. His first mandate was known as 'Revolution on the March' because of the favourable legislation towards the working class. It produced strong reactions among industrialists, landowners, the Catholic Church and the Conservative Party. With the exception of President Lopez's administration, the growing anti-labour sentiment is evident in Alberto Lleras Camargo's speech in 1945 in front of the National Association of Industrialists (ANDI):

> I am going to take advantage of this occasion to address from this platform the workers of Colombia to formulate a most zealous call to change, while it is still possible, the conduct that leads to the circumstance that when the worker, urban or rural, obtains better salaries, achieves the right to overtime pay, and receives protection against work hazards, illness, and unemployment: Production begins to diminish its rate, without any other cause than deficiencies in the realization of tasks ... This has to end ... what is guaranteed by the constitution is the right to work and not the right not to work. (República de Colombia 1946, cited in Oquist 1980: 235)

Liberals Against Conservatives or Capital Against Labour? Class Speaks Louder

In the twentieth century, there was also a conflict of an intra-class nature along party lines (Conservative versus Liberal) rather than elite divisions (such as landowners, industrialists and merchants). In the period 1930–40, Liberals and Conservatives engaged in a major power confrontation as each party attempted to construct political hegemony. In 1947, the tensions acquired a new intensity, which eventually led to the retirement of the Liberals from the first National Unity government (Oquist 1980). Increasing levels of violence between the parties led to the death of an estimated 14,000 in that year alone. During the same year, there was a general strike that led to waves of state repression (Pearce 1990). Between 1946 and 1966, Colombia was the scene of one of the most intense and protracted instances of widespread violence in the twentieth century. La Violencia,[19] as this civil war has been remembered in history, was of great intensity, widespread, lasted about 20 years and had a differential geographical distribution within Colombia (Oquist 1980).[20] It took 200,000 lives in the period between 1946 and 1966. 'Decapitations, mutilations, sex crimes, and with them, robbery and destruction of homes and land characterized La Violencia' (Pearce 1990: 54). Most historians date the eruption of La Violencia to the Bogotazo – the popular spontaneous uprising that took place in Bogotá in response to the assassination of the charismatic populist Liberal leader Jorge Eliecer Gaitan on 9 April 1948 during his presidential campaign. The initial death toll in the capital city was 2,585 people. This was followed by confrontations between Liberals and Conservatives throughout the country as the Conservative government and landowners' private armed groups, known as *pajaros*, unleashed violence against the Liberals in the countryside. The first Liberal guerrillas were created in 1949 by members of the Liberal Party and consisted mostly of peasants (Pearce 1990). In order to bring an end to La Violencia, the Conservatives and Liberals made a political pact in 1958, known as the National Front (Frente Nacional), which established that the presidency would alternate between the two parties for a period of 16 years and all positions in the three branches of government throughout the country would be distributed evenly between them (Safford and Palacios 2002). Nonetheless, La Violencia continued until 1966 mostly manifested in inter-class conflicts. The National Front barred the PCC from the conventional political process (Brittain 2010).

Despite its appearance as a battle between opposing political fronts, La Violencia was underlain by a strong inter-class conflict. According to Garcia (1971), it constituted the dismemberment of the popular movement in three stages:

1) an offensive against the most powerful unions and popular organizations by the provisional government of Alberto Lleras;
2) an annihilation of popular leadership, including the assassination of Gaitan; and
3) the establishment and consolidation of oligarchical support structures with the application of a formula of economic liberalism and political absolutism.

Another important point brought up by Oquist is that La Violencia corresponded to the ascendancy of finance capital and was marked by the highest rate of return on investments that Colombia had ever experienced: 'Rampant inflation redistributed income regressively while the economic advances that the working classes had achieved since the 1930s were reversed as the fascist policies of the government crushed the militant labour unions and their strikes' (1980: 134). Furthermore, it was also during La Violencia that the expansion of agribusinesses and cattle-ranching *latifundios* through the dispossession of peasants took place (Zuleta 2005). Liberal and Conservative landowners, frequently with the support of state armed forces, violently subdued peasant resistance whenever it challenged the existing power structures.

> The same elite that had been in charge before La Violencia, emerged in full control after it. The challenge from below had been defeated before it took hold ... The peasants meanwhile, rather than concentrate on their own class interests which might have turned them against the ruling elite, instead killed each other on its behalf. (Pearce 1990: 65)

Even though at one point Liberal guerrilla leaders came to fight alongside the armed peasant groups associated with the PCC, as Pearce points out, by 1953 most Liberal chiefs (who belonged to the capitalist class as opposed to the guerrilla body consisting of peasants) were against the Communists:

> Liberal peasant guerrillas began to break from the Liberal ranchers who had initially helped to create them. The landowners resented the guerrillas' demands for money and supplies, although they had happily used them to defend their property and interests. When the army began to move against the guerrillas, they put their property before their partisan loyalties and formed a special paramilitary group, the 'guerrillas of peace' to hunt down the peasant guerrillas. (Pearce 1990: 57)

Even if we consider the intra-class dimensions of La Violencia, it is important to see that it was not the political colours in themselves that generated the conflict, rather, the political power struggle, as Guillen (1963) argued, was over who would win the right to enrich themselves by managing the public treasury. 'Between 1946 and 1953 Colombia's Liberals and Conservatives killed each other by the thousands each year in a more than active scramble for economic rewards and appointments to governmental positions ... desire for economic rewards stemming from governmental control' (Guillen 1963, cited in Oquist 1980: 144).

The State's Iron Fist and the Birth of the FARC

As mentioned earlier, by the 1940s the Communist Party had supported the formation of organic peasant defensive structures against violence by the state and the landowners' private armed groups. These self-defence bodies had been joined by former Liberal peasant guerrillas who had broken away from their Liberal Party founders and allied with the Communists, partially motivated by the exclusionary nature of the Colombian political system during the National Front (Holmes, Gutierrez and Curtin 2008). In 1955, the PCC was declared illegal (Zuleta 2005).

In their areas of influence they [the guerrilla/PCC leadership] encouraged the peasant communities to share the land among the residents and created mechanisms for collective work and assistance to the individual exploitation of parcels of land and applied the movement's justice by collective decision of assemblies of the populace. These became areas with a new mentality and social and political proposals different from those offered by the regime. The decisive factor was the presence in power of the people themselves. (FARC-EP 1999: 15, cited in Brittain 2010: 8)

These self-defence Communist enclaves[21] were denounced by the Conservative Party Senator Alvaro Gomez Hurtado as 'independent republics' and accused of adhering to politics from Moscow (USSR) instead of the Colombian constitution. Consequently, the Conservative President Guillermo Leon Valencia vowed to exterminate these enclaves at any cost. On 27 May 1964 a series of heavily US-supported military operations were carried out by the Colombian state consisting of aerial bombardments and a 16,000 ground troop offensive against the settlements in the Marquetalia region in the southern part of the Departments of Tolima, Huila and Cauca. Those who managed to escape into the mountains formed the nucleus of what became the largest and longest-lasting armed insurgent movement in Latin America – the FARC, officially founded in 1966 (Brittain 2010).[22] Also towards the end of La Violencia other smaller guerrilla groups were formed, including the National Liberation Army (Ejercito de Liberacion Nacional, or ELN).[23]

'Weak' or Capitalist? The Evolution of the Colombian State

Comparing the twentieth-century Colombian state to its predecessors in the previous centuries, Oquist (1980) discerns a pattern of evolution. During colonial time the state was powerful and operated within a strong structure of social domination. In the nineteenth-century republic the state was weak but still based upon a strong class structure to which there were no serious sustained long-term challenges. The twentieth century, on the other hand, was characterized by a weakened structure of social domination and a process of a strengthening of the state.

The contemporary Colombian state was constructed during the 1920s and 1930s. There were two important forces that shaped and reinforced the roles the state assumed. The first was the sharp inter-class contradictions that had developed by that time, manifested in strikes, marches, land occupations, land disputes and the formation of large militant rural and urban social movements. The other was the control of the state by the ruling classes, especially through the *gremios* which ensured that state intervention would only take place to secure the interests of the capitalist class but not to address the needs of the working majority. Thus, the state emerged as an 'interventionist, non-pluralist entity that either absorbs or represses the social forces and organizations that are political actors' (Oquist 1980: 151). The Conservative and Liberal parties excluded any third party from the political order.[24] Hence, the principal functions of the state became to ensure the continuation of the existing class structure and to facilitate the capitalist development of the country. The Colombian state is a clear illustration of Wood's (1981) argument, grounded in Marxist theory, that the ultimate secret of production is a political one (see Chapter 2).

What some describe as the 'partial collapse of the state' (at the early stages of La Violencia) was followed by a rapid and sustained strengthening of the state, the central element of which was the increased importance of its coercive apparatus. The professionalization and modernization of the state military began in 1907 with the establishment of the National Military Academy (Oquist 1980). In 1945, the Colombian army numbered 8,000 men; by 1949 it had reached 20,000 men. Colombia was the only Latin American country to send troops into the Korean War in 1951. It was also one of the first five Latin American countries to sign a mutual defence assistance agreement with the US in 1952. In 1953, the military government of General Gustavo Rojas Pinilla created the Colombian Intelligence Service agency which in 1960 became the Administrative Department of Security (Departamento Administrativo de Seguridad, or DAS) – the country's central and most powerful intelligence institution. Between 1961 and 1967, Colombia received $60 million from the US in military assistance for counter-insurgency and economic development – all administered by the army – and a further $100 million in military equipment (Pearce 1990).

From 1902 to 1989, the use of violence by state forces as well as private armed forces associated with the various sectors of the capitalist class (especially against the peasant, indigenous and urban labour movement) became the predominant method of confronting challenges to the status quo. The expansion of the state's coercive apparatus took place simultaneously with two other developments. One was an increased reliance on state repression by the various sectors of the capitalist class. A legal mechanism that facilitated this was the notion of a 'state of siege' permitted by Article 121 of the Constitution, under the following circumstances:

> In the case of external war or internal disorder ... By such declaration, the government will have beyond normal legal authority, that which the Constitution authorizes for times of war and public disorder and that which, according to the accepted rules of international law, is applicable during war between nations. (Cited in Reyes 1991: 148)

Once a 'state of siege' is decreed, civilians are investigated and prosecuted by military tribunals and the government has the power to take any measures to counteract disturbances of public order. Between 1948 and 1984, a 'state of siege' was decreed in Colombia 15 times, for a total of 25 years and nine months. In other words, over a period of 36 years, Colombia had only ten years and three months of full juridical-institutional normality (Reyes 1991).

The second development in the strengthening of the state's coercive apparatus was a new strategy for enabling the state to more effectively defend the interests of the capitalist classes and confronting the insurgency challenge. It involved the creation of armed bodies outside the military institution which nonetheless worked with the state. Decree 2298, which was passed in 1965 and became Law 48 in 1968, laid the legal foundation for the establishment of paramilitarism by authorizing the executive branch to create civil patrols by decree and ordering the Ministry of Defence to supply them with weapons normally restricted to the exclusive use of the armed forces. Subsequently, paramilitary bodies called 'civil defence forces' were designed by and incorporated within the Colombian military system (Stokes 2005). In the 1980s, sectors of the capitalist class continued the paramilitary strategy by directly involving themselves in the creation of more paramilitary groups and expanding the existing ones. While the Colombian dominant classes had commonly relied on private armed groups since the nineteenth century (that is, the Conservative's *pajaros*), it was only in the second half of the twentieth century that an intense, long-term and well-coordinated relationship between these private armed groups and the Colombian state became an established practice. This was made possible through a legal framework that had set the ground for the design, training, equipping and administration by the state military of armed bodies outside its institution. It had such long-lasting effects that even when paramilitarism was outlawed in 1989, paramilitary groups continued to grow and multiply with the unofficial support (and often direct involvement) of the state.

Neoliberalism and Militarization: The Peace of the Rich is a War Against the Poor (1990–2013)[25]

Colombia's Open Veins in the Neoliberal Era[26]

During this period Colombia's economy became increasingly integrated into the international division of labour and globalized production chains (O'Connor and Bohorquez 2010) through a continuation of agricultural exports such as coffee, bananas, cotton and beef (including an increase in some new ones such as palm oil, flowers and seafood), as well as an increase in natural resource extraction (particularly in emeralds, oil, coal and ferronickel). The illegal drug trade also continued and witnessed the formation of new elaborate networks following the Medellín Cartel of the 1980s, such as the Cali Cartel and the Norte del Valle Cartel.

The behaviour of the Colombian state in the 1990s is consistent with the description that Robinson (2004) offers with regard to the responsibilities of the TNS in serving global capital accumulation mainly through fiscal and monetary policies, basic infrastructure necessary for global economic activity, and coercive and ideological apparatuses for social control. Beginning in 1990, and influenced by the Washington Consensus,[27] the government of President Cesar Gaviria Trujillo (1990–94) began a comprehensive neoliberal restructuring known as the 'Economic Opening' (*Apertura Economica*). It included several components. One was financial liberalization, which allowed the free flow of foreign exchange and international capital and the abolition of exchange rate controls over foreign investment. Another was trade liberalization that eliminated most non-tariff barriers and dramatically reduced tariffs. The *Apertura* also included labour deregulation consisting of modified labour laws that increased the flexibility in hiring and dismissing workers, as well as waves of privatization in banking and telecommunication. The administration of President Ernesto Samper (1994–98) continued the neoliberal agenda. After one decade of neoliberal policies, unemployment had reached 20 per cent in 2000 (Holmes, Gutierrez and Curtin 2008). Between 1995 and 2000, Colombia's ranking on the UNHD Index declined and overall poverty increased from 55 per cent to 60 per cent (Yepes 2002, cited in Aviles 2006). Starting in 2000, in addition to agribusiness, the other predominant sector of Colombia's rural political economy became mineral resource extraction. In 2001, the Colombian Congress approved Mining Code Law 685, which erased restrictions on foreign ownership of concessions, liberalized the cadastral system, and removed restrictions on licensed corporate mining activity on public lands. The law also required authorization for artisanal mining on state properties,[28] which led to the expropriation of independent artisanal miners by the state or paramilitary forces from territories which are claimed by licensed companies (including MNCs) (O'Connor and Bohorquez 2010).

Between 2002 and 2010, under President Uribe, Colombia experienced a wide range of neoliberal reforms that furthered the developments initiated by previous administrations. A thorough privatization of public resources and service providers, along with a significant downsizing of national enterprises, took place under pressure from the IMF in exchange for a $2.1 billion loan. Uribe's government privatized Bancafe (one of the largest banks[29]), Telecom (the public telecommunications company) and Minercol (the state entity responsible for administrating the extraction of coal and mineral resources including gold, silver, platinum, nickel and emeralds) (Leech 2003; O'Connor and Bohorquez 2010). ECOPETROL (Colombian Petroleum Company, the state-owned oil company) was liquidated (Leech 2003). Some of the country's largest public hospitals were closed down. Tens of thousands of full-time unionized workers lost their jobs. Public-sector workers were laid off, pension programmes were reformed, and the royalties that foreign oil corporations were required to pay Colombia were lowered.

President Uribe's government also signed free trade agreements with the US in 2006 and with Canada in 2008. These agreements consolidated Colombia's position in a division of labour dating back to colonial time, as a seller of primary resources (agriculture and minerals) and a buyer of manufactured goods and high technology commodities. In addition to dependency on stable access to Northern markets (Grinspun 2003), the further expansion in agricultural exports (encouraged by the agreements) means that less and less land is available for growing staples for local consumption, thus threatening the nation's food autonomy. Clearly, the poor are most likely to be affected in this case. President Santos has not challenged the two pillars of his predecessor's agenda: free market and investors' confidence on the one hand, and security on the other, even though he has tried to mask aggressive capitalist investor-friendly policies with a socially progressive discourse, such as the Victims and Land Restitution Law of 2011 (a discussion of which is offered in the next chapter) and the 'National Development Plan'. In reality, the latter focuses on ways to create even more favourable conditions for large-scale and foreign mining operations, on the agroindustrial model of agriculture (at the expense of driving small-scale farming into extinction), on the construction of infrastructure for megaprojects which are most of the time destructive to local livelihoods, as well as on privatizing education and health care. One concrete illustration of this can be found in the mining sector. Currently around 40 per cent of the national territory is used for mining and energy exploitation. In the Department of Cauca, 84 per cent of the territory has been handed over to foreign investors (Zamora 2013). Many of the natural resource exports from Colombia are done through US or European-based MNCs. For instance, all of the exports of ferronickel are done through the UK company BHP Billiton. The latter pays Colombia only 10 per cent of the total revenue from selling Colombia's nickel abroad (Corriente Marxista 2009). Similarly to primitive accumulation processes during colonial times, the detrimental consequences of mining and energy megaprojects under the Santos administration continue to be felt the most by the 102 indigenous groups and the more than four million Afro-Colombians (Zamora 2013), thus reproducing the interplay of class and racial inequalities.

As with many other Latin American countries, Colombia's experience since 1990 is an illustration of how neoliberalism intensifies the processes through which economic and political power becomes concentrated in a few hands and resources are transferred from the public to the private sphere. Neoliberal restructuring guarantees huge profit opportunities to the already economically powerful groups, such as local elites, MNCs and financial operators, while excluding the participation of the low-income population (Lefeber 2003). Mechanisms external to the country, such as trade laws and the external debt, reinforce Colombia's subordinate position within the global hierarchy of power, while internal mechanisms in the form of macroeconomic policies contribute to the marginalization of the low-income population. In 2010, 45.5 per cent of Colombians lived in poverty (UNDP 2011), and 13 per cent lived in extreme poverty (Prensa Latina 2012). Close to 20 per cent of the population were homeless (DANE 2009). Approximately 15 per cent of Colombian children were malnourished (Holmes, Gutierrez and Curtin 2008).

One of the significant characteristics of Colombian society that became pronounced in the 1990s is what Richani (2010) calls a 'reactionary class configuration'. The latter comprises the capitalist class associated with the export economy (both legal and illegal) and the old landed elite, both united by a common enemy – the insurgency. Another important feature of the class structure in the last 20 years has been the growth of a specific sector of the working class: an 'underclass of supernumeraries or "redundants" who are alienated and not absorbed into the global capitalist economy and who are structurally under- and unemployed' (Robinson 2004: 23). In Colombia, these are informal labourers such as street vendors, recyclers, street sex workers, shoe shiners and some domestic workers, to mention only a few. Many of these are rural migrants who have been directly or indirectly forced to leave their land and find themselves living in squatter settlements. Around 45 per cent of Colombians work in the informal economy unprotected by labour laws (Prensa Latina 2012). The aggravation of the existing patterns of inequality present in the previous epochs has led to a sharpening of the prevailing class contradictions.

The 'Weak' State Tightens its Grip

The relationship between the concentration of wealth and the use of violence has become more important than ever during the epoch of neoliberal restructuring. The growth in agribusinesses and mineral resource extraction activities by local and foreign companies, as well as the investment of drug capital in high quality land, have greatly accelerated the counter-agrarian reform across the entire country and have produced a human rights disaster of emergency proportions. Close to six million people have been forcibly displaced (CODHES 2012), overwhelmingly through terror-based strategies, and land titles for millions of hectares of land have been transferred into the hands of the elite. Parallel to these agrarian conflicts, a strategy of annihilation targeting labour unions has been underway, aimed at de-unionizing the Colombian labour force and ensuring a pool of precarious workers willing to work at any price in order to survive. The result has been a drop in the percentage of unionized workers from 12 per cent in 1988 to a mere 4 per cent in 2009 and over 3,500[30] unionists dead between 1990 and 2005 (ICFTU 2005b). These assaults against rural and urban labour have been enacted through a combination of violence-based methods (threats, assassinations, forced disappearances, torture and sexual violence carried out by paramilitary and state forces) as well as legal mechanisms.

Legislation has an important role in empowering state agents to perform their job in ways that not only breach civil liberties and privacy, which are commonly thought of as key ingredients in the fabric of democracy, but also violate basic human rights. In this context, legal mechanisms operate in three interrelated ways:

1)   authorizing invasive forms of social control through the increase and diversification of surveillance practices;
2)   allowing the police and military to use excessive force, thus curtailing civil rights and liberties; and
3)   providing schemes through which dissent can be criminalized.

One of the best illustrations of the first method is the Defence and Democratic Security Program, put into effect by former President Uribe in 2002.[31] Another example of invasive surveillance practices and arbitrary types of interference was DAS's[32] systematic engagement between 2004 and 2009 in illegal phone tapping, email interception, and the monitoring of the activities of labour unionists, human rights activists, journalists writing about the state's connections to the paramilitary, left-wing politicians, as well as magistrates and Supreme Court justices investigating former paramilitaries and politicians linked to the paramilitary (HRW 2012). Alfredo Correa de Aldreis, a sociologist, was one of the many individuals being followed by DAS. He was later murdered in the city of Barranquilla.

Legislation also serves to criminalize dissent and social protest while justifying repressive tactics by state agents. The labelling of activists and social movements as guerrillas is the principal way in which their demands are discredited and their actions criminalized. Mobilizations and marches by trade unions and student movements have been especially harshly affected. In 2004, the Ministry of Social Protection declared the majority of strikes to be illegal, often citing public order as the main reason. This and other recent changes in labour law have made it harder for workers to exercise their freedom of collective bargaining and association (Moloney 2005). Arbitrary legal proceedings relating to charges of subversion against trade unionists and other human rights defenders have been common. On numerous occasions trade unionists have been killed whilst under criminal investigation or shortly after charges against them were dropped. Many of the criminal proceedings opened against trade unionists have been initiated on the basis of accusations by the armed forces and paid informers instead of on evidence gathered in the course of independent and impartial criminal investigations by the civilian investigative authorities (AI 2007). The excessive prolongation of pre-trial detention and criminal proceedings against those accused by the state is another way in which legal mechanisms are put to work. Liliani Obando, a Colombian academic, trade unionist, human rights defender, sociologist and filmmaker, was detained in August 2008 and kept in preventive detention for three years and six months in Buen Pastor Prison, Bogotá, without being convicted of any crime. At the time of her arrest she was working on her dissertation on the Federation of Agricultural Workers and Small-Scale Farmers (Federación National Sindical Unitaria Agropecuaria, or FENSUAGRO). She had been falsely accused of rebellion and raising funds for FARC based on evidence found among the files in laptops belonging to former FARC commander Raul Reyes which were recovered after the bombing-raid and ground assault on the FARC encampment in Ecuador on 1 March 2008. In its 2008 report on these files, Interpol stated that the computer documents had been kept for one week by the Colombian authorities before being handed over to Interpol. During that week, the Colombian authorities actually modified 9,440 files, and deleted 2,905, according to Interpol's detailed forensic report (Grandin and Salas 2011).

While legal mechanisms furnish the state's coercive apparatus with legitimacy and credibility in targeting those who express opposition to the status quo, the most common method by which the state affects and controls the civilian population is armed force, expressed in state-sanctioned violence and militarization. In Colombia, state-sanctioned violence is manifested in arbitrary arrests, house raids, excessive use of force at public protests and demonstrations, aerial fumigation that destroys food crops and livestock, indiscriminate bombing, intimidation, beatings, torture, forced disappearances, extra-judicial executions, and coordination and collaboration among the state armed forces and paramilitary groups in carrying out any of the activities listed above or any other forms of human rights violations (Hristov 2009b). As Miliband argues, the military and police in capitalist countries are the coercive agents of the existing social order and they are ready 'to take the field, so to speak, against striking workmen, left-wing political activists, and other such disturbers of the status quo' (1973: 123). For instance, Colombian small-scale farmer leaders who participate in demonstrations after negotiations with the government have often been labelled subversive and have been the victims of serious human rights violations carried out by the security forces. In August 2003, Hermes Vallejo Jiménez, a peasant leader in the Tolima Department and one of the founders of the Association of Small and Medium-Scale Farmers of Tolima (Asociación de Pequeños y Medianos Agricultores del Tolima, or ASOPEMA), was detained along with 25 others, including two members of the National Association of Hospital and Clinic Workers (Asociación Nacional de Trabajadores Hospitalarios y de Clínicas, or ANTHOC) based on claims made by informants found by Grupos de Acción Unificada por la Libertad Personal (GAULA) who labelled the detained as members of the ELN guerrilla. In May 2006, members of the GAULA detained FENSUAGRO leader Miguel Angel Bobadilla and his partner Nieves Mayusa. During the operation, the authorities attempted to force the couple's elder child to say that her parents were FARC guerrillas (AI 2007).

In March 2011, the Seccion de Investigaciónes Judiciales e Inteligencia (SIJIN) detained Jose Antonio Toroca, the President of the Communal Action Board in the rural area Flor Amarillo in the municipality of Tame, Department of Arauca, who had led protests against the crimes committed by the armed forces against children. Most recently, he had been very outspoken about a case where two little girls were sexually abused and one of them killed in the presence of her baby brothers by members of the armed forces (Humanidad Vigente 2011).

The Colombian state's coercive apparatus has experienced significant militarization since 1990 expressed in 1) the increased presence of the state troops in certain areas; 2) the increased control of certain territories and their inhabitants by the army (that is, random searches of civilians conducted by military personnel at roadblocks or other public spaces); 3) the increase in the number of military bases; and 4) the increase in the number of state troops (Hristov 2009b). The War on Drugs and the War on Terror served as an impetus for the strengthening of the state's coercive apparatus both quantitatively and qualitatively. The US has played a fundamental role in the militarization of Colombia through funding, training and the provision of equipment, troops, and experts,[33] while disregarding the fact the Colombian armed forces have one of the worst human rights records in the world. As Miliband put it, the purpose of US foreign policy has always been to prevent the emergence of regimes fundamentally opposed to capitalist enterprise: 'As a general rule, the American government's attitude to governments in the Third World ... depends very largely on the degree to which these governments favour American free enterprise' (1973: 78). Since the Second World War, under the responsibility of the Southern Command of the Pentagon, the US has maintained a number of military bases throughout Latin America, which has grown since 11 September 2001 (Marion 2002). US–Colombian military cooperation has been governed by conditions set in a number of bilateral agreements, including the 1952 Mutual Defence Assistance Agreement, the 1962 General Agreement for Economic, Technical and Related Assistance, and related subsequent agreements in 1974, 2000 and 2004 (US Department of State 2009). US assistance has gone into combating subversion, narco-trafficking, terrorism and other perceived threats to Colombia's national security (Torrijos 2010) through numerous large and well-funded security missions and projects.

Starting in the 1980s, the Andean region has been the main focus of the US War on Drugs. Between 1990 and 1994, $2.2 billion in mostly military assistance was allocated to Colombia, Peru, Bolivia and Ecuador. In Colombia in particular, funds were used for the creation of new police squadrons and massive fumigation operations for the eradication of illegal crops. In 1999, the Plan for Peace, Prosperity and the Strengthening of the State (Plan para la Paz, la Prosperidad y el Fortalecimiento del Estado), known as Plan Colombia,[34] was put in place by the administration under Colombian President Pastrana and US President Clinton. This was a six-year strategy (1999–2005) for eliminating drug-trafficking and organized crime, strengthening institutions, and promoting social and economic development. The total cost of Plan Colombia was $10 billion – 35 per cent of which was financed by the US with the rest covered by Colombia. Of this total amount, 26.5 per cent was devoted to the strengthening of institutions, 16 per cent to social and economic development, and 57.5 per cent to counter-narcotics operations and combating organized crime. In practice, the emphasis was on militarized counter-narcotics operations (mostly focused on targeting the primary producers – the *campesino* coca growers) and intensive counter-insurgency operations, both conveniently combined as a response to the 'narco-terrorist threat' (Rojas 2007). The increased militarization that has been made possible through

Plan Colombia has also served to provide security for foreign investors throughout the countryside. As far as reducing the cultivation and trafficking of illegal drugs, the results have been quite disappointing. Land under coca cultivation increased from 136,200 hectares in 2000 to 157,200 hectares in 2006. Moreover, the United Nations World Drug Report 2008 estimated that there had been a 27 per cent increase in the area cultivated with coca in the period 2006–7. Colombia remains one of the biggest producers and exporters of cocaine in the world (Acevedo 2008).

At the beginning of 2007, Uribe's administration presented the Strategy for Strengthening Democracy and Social Development 2007–13 (Estrategia de Fortalecimientode la Democracia y del Desarrollo Social), which became known as Plan Colombia II. The objective was to consolidate the achievements of the first phase of Plan Colombia and the Defence and Democratic Security Program. Plan Colombia II has six components: the war against narco-trafficking and terrorism; strengthening of justice and human rights; internationalization of the economy; social programmes; attention to displaced people; demobilization, rendition and reintegration. The total cost for Plan Colombia II was estimated at $43 billion. A mere $1.7 billion was intended for socio-economic development (Rojas 2007). The main focus continues to be on narco-trafficking, organized crime and insurgency, together labelled as 'narco-terrorism'. Between 1998 and 2008, under Plan Colombia I and II, US aid to Colombia included the following allocations: military and police aid ($5.5 billion); economic and social aid ($1.2 billion); military equipment ($1.3 billion); counter drug operations ($176 million); and humanitarian and civic assistance ($871,975) (Acevedo 2008). In her evaluation of Plan Colombia I, an analyst from the Beckley Foundation Drug Policy Programme writes:

> Plan Colombia demonstrates the negative consequences of combining the war on drugs and the war against terrorism ... Plan Colombia has shown how addressing the originally discrete, yet what have grown to be inter-related, problems of violent 'Revolutionary' politics and the illicit drugs trade via a single strategy can have deleterious consequences. As such, the misleading association between drug-trafficking and insurgence represented in the politically popular phrase 'narco-terrorist threat' needs to be reconsidered and separated ... [I]t is clear that a military approach against drug trafficking fails to achieve democratic stability and peace since the increasing militarization of the anti-drugs efforts can have devastating effects in terms of displaced population,

intensification of the conflict and the escalation of the violence. (Acevedo 2008: 10)

In addition to financial aid for specific projects, US military involvement in Colombia is expressed in intelligence operations, maintenance of bases, training programmes for local troops, and provision of equipment such as radar installations, helicopters and ammunition. In fact, out of the non-NATO countries, Colombia is second after South Korea in terms of the number of US-trained military personnel present in the country (CIP 2003). During the administration of President George W. Bush, the US intensified military and economic bilateral ties with Colombia, including the definition of a free trade agreement and the signing of the Supplemental Agreement for Cooperation and Technical Assistance in Defence and Security (SACTA) in November 2009 (Torrijos 2009). The defence agreement gives the US military access, use, and free movement among two air bases, two naval bases and three army bases, in addition to an existing two military bases, as well as all international civilian airports across the country. It also grants US personnel full diplomatic immunity for any human rights abuses or other crimes committed on Colombian soil (Janicke 2009). The Colombian Constitutional Court suspended the agreement on the grounds that it was never approved by the Colombian Congress and thus was unconstitutional. The US military forces situated across Colombia had to withdraw to other Colombian bases until the Colombian Congress approved the agreement in a democratic manner (Pitarque 2010). These other bases are located in San Andres Island, Miraflores, Ariquita, Santa Marta and Puerto Asis (Castro 2002).

In May 2004 the Patriot Plan (*Plan Patriota*) was announced. This was a joint initiative put forward by the Colombian and US governments for a massive military offensive, including the deployment of 15,000 troops into territories occupied by the FARC. Moreover, the Patriot Plan provides legal privileges to US private companies and their mercenaries (Calvo 2004).

In 2010, the Colombian Ministry of Defence launched the Plan Salto Estrategico with the objective of consolidating territories, such as Catatumbo, Cauca, los Montes de María, the southern part of the Department of Tolima and the Departments of Nariño and Arauca, which have a high presence of guerrilla and other illegal armed groups. According to the government, this project has six steps:

1) identify the territories for military operations;

2) assess the degree of presence of guerrilla or other leaders who can be captured, killed or made to demobilize;
3) capture the maximum number of combatants and subject them to judicial proceedings;
4) conduct anti-narcotic operations;
5) occupy the areas which will likely be sought as refuge by those who manage to escape; and
6) engage in social recovery of the territories (Semana 2009d).

On 29 March 2012, US Joint Chiefs of Staff Commander General Martin Dempsey announced the newly formed Vulcan Joint Task Force in Colombia, led by Brigadier General Marcolino Tamayo Tamayo, which consists of 10,000 soldiers, three mobile brigades and one fixed brigade, operating from a base in Tibú, in the Catatumbo region Department of Norte de Santander, just two miles from the Venezuelan border. Once again, the objective of this new task force is to fight narco-trafficking and the insurgency as in the case of other similar joint task forces that have been created in Tumaco (Department of Nariño), Miranda (Department of Cauca) and Tame (Department of Arauca). Dempsey announced that the US will send to Colombia brigade commanders with practical experience in Afghanistan and Iraq to work with the Colombian police and army combat units that will be deployed in areas controlled by the rebels. Colombia has already formed special commando units for hunt-and-kill missions as part of an aggressive military campaign through which the government hopes to reach its goal of reducing the FARC guerrillas by 50 per cent in two years (Pimiento and Poland 2012).

On 15 April 2012 presidents Obama and Santos met during the Americas Summit and agreed on a new military regional action plan that will include training police forces in Central America and beyond (discussed in Chapter 1). The expansion of counter-insurgency forces in Latin America is also part of a new national security strategy released by the White House in February 2012 which, according to US Defence Secretary Leon Panneta, will introduce 'innovative methods' for supporting counter-terrorist forces and expanding US influence in the region (Pimiento and Poland 2012). From a military point of view, Colombia is a very important Latin American ally for the US because of its strategic location in relation to the rest of Latin America. This is especially so given the rise to power of anti-neoliberal governments in Venezuela, Bolivia and Ecuador, and their emphasis on national sovereignty and autonomy from US imperialism. Thus, the militarization of Colombia not only serves to contain the insurgency and provide security for US enterprises operating inside Colombia, but also to maintain a level of preparedness to deal with any challenges that might arise from possible alliances between US-unfriendly governments in the region and social movements inside Colombia.

Paramilitarism as the Enabler of Economic Domination and Political Rule (1990–2006)

As noted above, the legal foundation for the establishment of paramilitarism was laid by Decree 2298, which was passed in 1965. Converted into law in 1968, and remaining in effect until 1989, it authorized the executive branch to create civil patrols by decree and ordered the Ministry of Defence to supply them with weapons normally restricted to the exclusive use of the state military. Subsequently, paramilitary bodies called 'civil defence forces' were designed by and incorporated within the Colombian military system (Stokes 2005). This was part of the military project Plan Lazo, launched by the Colombian and US administrations, designed to defeat the guerrilla and also eliminate the potential for subversion by targeting sectors of the civilian population that were considered fertile ground for Communist indoctrination and guerrilla support. The target for paramilitary actions was the 'internal enemy', which extended beyond armed insurgents and encompassed legal political organizations demanding real democratic reforms, educators, leaders of dissident groups, peasant movements and labour unions.

Starting in the 1980s, the capitalist class of Colombia played a more direct role in the setting up of paramilitary bodies, and used the same term as the state had given them in the 1960s – 'self-defence' forces. Nevertheless, the state promoted and supported these initiatives and also at times participated directly. The self-defence groups, as will be illustrated, were very similar in nature to the original paramilitary groups initiated earlier by the state. This second wave was enacted by large-scale landowners, cattle-ranchers (Department of Cordoba), agribusiness owners (such as banana plantations in the Department of Magdalena and the Uraba region), the mining entrepreneurs (particularly those in the emerald business in Department of Boyaca), and drug-lords. A key factor that facilitated this process was the legal framework established in the 1960s, which paved the way for non-state armed groups to attain legitimacy. The alliances among sectors of the capitalist classes, politicians and the military gave rise to the emergence and territorial expansion of various groups throughout different parts of the country, accompanied by a progressive growth in their financial and military strength, despite the outlawing of paramilitarism in 1989.[35] Towards the end of 1994, under the leadership of Carlos Castaño, the first National Conference of Self-Defence Groups was completed and three years later the United Self-Defence of Colombia (AUC) was formed.[36] The period between 1997 and 2003 was a critical one for the growth of the AUC and witnessed the emergence of the paramilitary groups Bloque Norte, Catatumbo, Centauros, Vencedores de Arauca, Calima, Pacifico, Cacique Nutibara and Capital Central Bolivar. Although the AUC comprised groups that differed in their specific interests and objectives, it had a plan and strategies for expanding its influence and domination at a national level.

Beyond Horror: Cannibalism and Crematoriums

Paramilitary groups typically attack any social forces that emerge as an obstacle or a challenge to the interests of local and foreign capitalists. Their targets are trade unions, indigenous, women, youth and peasant organizations, educators, intellectuals, journalists, students and human rights activists. In my book *Blood and Capital*, I provide detailed coverage of numerous cases that illustrated the various forms of human rights violations carried out by the paramilitary against each of these categories of people, jointly perceived as 'the internal enemy'. Here, I provide a glimpse into paramilitary atrocities, followed by a thorough analysis of the complex amalgamation of violent and legal mechanisms that enable paramilitary operations to generate capital accumulation. Between 1993 and 2006, the paramilitary carried out 1,528 massacres that left 8,449 people dead (Semana Multimedia 2012). As Richani explains, 'The assassination of unionists and peasants is a ... concerted attempt to stamp out collective bargaining and unionization and hence to reduce both the cost and power of labour' (2007: 413).

It is difficult to exaggerate the horrifying nature of paramilitary slaughter. Two major considerations influence the way in which the paramilitary chooses to execute its acts of violence. One is fostering a culture of fear. For these armed groups, one of the prerequisites for establishing dominion over a territory and its people, is for the paramilitary group to be feared. This is accomplished by tactics such as torturing individuals in

Table 3.1   Percentage of Civilian Deaths and Forced Disappearances for which the Paramilitary was Responsible

Year	Percentage
1993	18
1995	46
1997	69
1998	76
1999	78
2000	79

Source: CCJ cited in Aviles (2006).

Table 3.2   Number of Unionists Assassinated 2001–2006

Year	Number of Unionists Assassinated
2001	209
2002	192
2003	102
2004	99*
2005	73
2006	77

Source: USLEAP (2011).
*Source: ENS (2010).

the presence of others, leaving corpses grossly mutilated, and engaging in practices that lead the victims to experience a painful death. The following is a testimony of a person forcibly displaced by paramilitary terror who was interviewed by Amnesty International in 2003:

> A stick was pushed into the private parts of an 18-year-old pregnant girl and it appeared through [the abdomen]. She was torn apart ... They [army-backed paramilitaries] stripped the women and made them dance in front of their husbands. Several were raped. You could hear the screams coming from a ranch near El Salado [Department of Bolívar]... (AI 2004)

Another factor that influences the way the paramilitary kills its victims and what it does with them subsequently is the need to keep the victim's whereabouts unknown. For as long as the victim's remains are not found and they are thus registered by the authorities as disappeared, the perpetrators cannot be convicted of murder. The paramilitary's answer to this was the creation of various crematoriums. For instance, groups led by Ivan Laverde, alias El Iguano, and Rafael Mejia, alias Hernan, operating in the Department of Norte de Santander, burned over 200 victims between 2001 and 2003 in such crematoriums. One witness from Villa del Rosario tells how this was done: 'It smelled like the devil ... They [the paramilitary] would first leave the corpses in the sun to dry up and when all that remained was bones, they would place them in the ovens.' Another witness adds: 'They would kill people, bury them in mass graves and after six months dig them up and put them in the oven. Other times, they would open up the corpses, remove all the internal organs, chop them up and then put them in the oven' (Rebelion 2009: 2). Confessions made by paramilitary chiefs Ivan Laverde, Rafael Mejia and Salvatore Mancuso before prosecutors under the Justice and Peace Law in April 2009 confirmed these stories (Rebelion 2009). Another unbelievably repulsive feature of paramilitary violence was the practice of cannibalism. Former combatants from the groups led by paramilitary chief Hector Buitrago and his sons Martin and Caballo testify that the tasks of chopping up human corpses, drinking the blood, frying the human flesh and eating it, were part of the intense and harsh training that new recruits had to undergo. The victims of this carnage were not only individuals killed as part of paramilitary operations, but also new recruits who did not pass the test at the end of each training session, as well as other combatants who had been left disabled by injuries they suffered during combat (Semana 2012c).

The Uribe administration opened formal negotiations with the AUC in July 2003 and in February 2006 the government announced the completion of the demobilization of the AUC. The rest of this chapter illustrates the paramilitary's links to the state institutions and the blending of institutional and parainstitutional mechanisms as a tactic of forced displacement. The next chapter reveals evidence of the continuation of paramilitarism and its connections to local and foreign enterprises as well as state institutions in the post-demobilization era.

Legalizing Land Theft: Anatomy of State–AUC Complicity in Forced Dispossession

While paramilitary operations generate wealth in a wide variety of ways (see Table 3.3), one of their principal activities is the forced displacement of populations from areas of strategic economic and/or military importance such as: land high in fertility; territories containing valuable natural resources such as minerals, gold, oil or precious woods; areas used by the guerrilla as transportation passageways; and fields with or suitable for potential illicit crop cultivation (Hristov 2009b).

Table 3.3   Sources of Funding for the Paramilitary

Dispossession	Services	Illegal Economic Activities	Legal Activities
– Illegal appropriation of land by causing the forced abandonment of rural properties.	– Protection for large estate owners against incursion and taxation by the guerrilla – Social cleansing (limpieza social) – Repression of social movements (for example, labour unions) through threats, murder and other forms of human rights violations – Facilitating the electoral victory of paramilitary friendly politicians by eliminating opponents and coercing the population.	– Drug-trafficking – Arms-trafficking – Oficinas de Cobro(criminal structures offering services for debt collection, assassination, and dispute resolution) – Robbery and theft – Extortion of landowners, small to medium-scale farmers, food producers and vendors, gas stations, transportation businesses, prostitution establishments, politicians, lottery-winners – Embezzlement of government funds.	– Land ownership – Plantations with palm oil and other cash-crops – Casinos, lotteries – Gas Stations – Shopping centres – Prostitution establishments.

Before discussing the dynamics and consequences of forced displacement, it is necessary to examine the conceptual significance of the latter and its relationship to the concept 'dispossession'.[37] Forced displacement has been defined in this work as cases where people have been driven by means of violence or threats to abandon their place of residence out of fear for their own safety or that of their family. What happens to the land once its owner has been forcibly displaced? Forced displacement is an act which results in the forced abandonment of property. Abandonment implies the temporary or permanent interruption of the use, enjoyment, access to and possession of things. Thus, forced abandonment is the result of the use of violence and displacement of the population (Comisión Nacional 2009). Most often the goal of forced displacement is forced abandonment of the property and consequently dispossession.[38] The latter is defined by the Project for the Protection of Lands and Heritage of the Displaced Population as:

> the action through which a person is deprived of his/her property, possession, occupation, or any other right he/she exercises over his/her property whether by judicial process, administrative act or through circumstances related to the armed conflict ... The intention of dispossession is the theft or expropriation of a good or a right. (Comisión Nacional 2009: 24)

Paramilitaries have been largely responsible for most of the forced displacements. A human rights report by Colombia's Ministry of Defence shows that the paramilitary has directly caused 46 per cent of the forced uprooting of people from their homes, compared to the guerrilla which is deemed guilty of 12 per cent (Contreras 2002). It is important to recognize that land in Colombia is a precious resource. Given the few opportunities for formal employment, land continues to be a vital source of subsistence for the rural population. At the same time it has been a source of economic and political power and thus, as the country's history has shown, land has been at the heart of the armed conflict. Since the 1990s, through the use of various terror strategies, the paramilitary has been massively displacing populations across the Departments of Antioquia, Córdoba, Valle del Cauca, Nariño, Caldas, Chocó and Bolívar, paving the way for the emergence of new agroindustrial and mining projects.

I will illustrate the manner in which forced displacement, forced abandonment, and dispossession have commonly taken place in Colombia. But first it will be helpful to give some background on the circumstances surrounding the property ownership of the small-scale farmers prior to their forced displacement. In Colombia, usually such farmers had obtained titles to land in one of two ways. In the first case, they had been allocated a parcel and had received a loan, as part of a government programme, to start their own farm. Legally, they could not sell the land in the first 15 years after the grant. Sometimes land grants had been made through collective title where a number of families had all been given one large farm. In the second case, the small-scale farmers had colonized a *baldío* (an empty, unused, or abandoned terrain) and after residing on it and farming it continuously for ten years had obtained a legal title. In such cases it was also common to be given a loan by the agrarian bank to buy initial equipment and seeds. In both cases, the crops cultivated on these small-scale farms were typically subsistence crops, such as beans, rice, yucca, corn, plantain and potato, which the farmers used to meet the food requirements of their own families and to sell the rest at the local market.

Typically, when paramilitary forces engage in forced displacement, their goal is to displace all residents in a given area so that many individual neighbouring plots are vacated, and the resulting total area can be converted into a single large estate. This bears an astounding resemblance to Marx's account of primitive accumulation in England: 'In several parishes of Hertfordshire, writes one indignant person, twenty-four farms, numbering on the average 50–150 acres, have been melted up into three farms' (1867/1990: 886). The first and least violent way in which many *campesinos* have been displaced is by being pressured through intimidation and threats into selling their land at unreasonably low prices. Often the coerced sellers do not even receive the full price they had been promised. In such cases, despite the fact that the sale was involuntary, it is still registered as a legal economic transaction between a buyer and a seller and is not regarded as involving any forced abandonment. Therefore, from a legal point of view, this act does not constitute dispossession, even though it constitutes forced displacement.

Under a different approach, when armed actors force people to abandon (but not sell) their property, several mechanisms come into play that ultimately turn the forced abandonment into a dispossession where the land is ultimately appropriated by the paramilitary. In the first type of case, the owner is coerced into signing a document stating that they agree to abandon their property. Alternatively, the signature is forged. Either way, INCORA[39] declares the land abandoned and it becomes state possession. The land is then 'sold' to the paramilitary by INCORA/INCODER at a giveaway price. Subsequently, the new owners improve the infrastructure and guarantee security by having cleared the area of small-scale farmers considered to be a social support base for the guerrilla. At that point they are able to sell the property at high prices to Colombian and foreign investors, cattle-ranchers,[40] cash-crop plantation owners, mining entrepreneurs and tourism businesses. In the Department of Cesar, for instance, before 2000 (when the paramilitary came to dominate the area) the price of one hectare of land was two million pesos. By 2009, the price was 60 million (Semana 2009c). Similarly, between 2005 and 2009 more than 70,000 hectares in Los Montes de María, Department of Bolívar, were sold to large investors at prices under 500,000 pesos. Today these same lands cost close to three million pesos (Semana 2010c).

A brief background on the institutions that serve as the crucial link in the process of legalizing ownership over illegally appropriated land is in order here. INCODER was created by the Ministry of Agriculture and Rural Development in 2003 by Decree 1300 and subsequently reformed by Law 1152 of 2007 (Ortiz 2009). This entity replaced INCORA. INCODER was officially expected to develop and promote different kinds of development programmes for rural Colombia by providing financial and technical support for agricultural, forest and fishing activities, facilitating access to land for small and medium-scale agricultural producers, promoting research and capacitation, and administering land acquisition and grants (INCODER 2012). However, ever since its inception it has been involved in the embezzlement of funds. During its first six years of operation, five of its managers were implicated in corruption scandals and dismissed. In 2008, INCODER received a budget of 32 billion pesos to assist 1,200 families in need. Only 79 families have benefited and the rest of the money has still been unaccounted for (Ortiz 2009). Most importantly, INCORA and subsequently INCODER have not been characterized merely by isolated cases of corruption, but have rather served to legalize the appropriation of land through violence and human rights violations. INCORA and INCODER have performed the most crucial function in the process between the moment forced displacement occurs and the moment of paramilitary land appropriation by: first, declaring the land abandoned (when in fact its inhabitants were coerced into abandoning it and fleeing); second, revoking the land title; and third, assigning the land title to paramilitary commanders, family members, their companies and *testaferros*.[41] In some cases INCORA has also heavily subsidized new farms (formed by joining the individual plots of illegally appropriated land), such as Asoprobeba in the Uraba region of the Department of Antioquia, formed by members of the Cattle-Ranchers' Fund of Cordoba and the paramilitary.

The following specific examples illustrated the mechanisms of dispossession at play. In 1991, 72 landless families were granted a collective title to the farm La Pola by INCORA in Chivolo, Department of Magdalena. In 1997, a paramilitary unit under the commander Rodrígo Tovar Pupo, alias Jorge 40, invaded La Pola and a few other rural areas with collective farms, including Pueblo Nuevo, San Angel, Chivolo and La Estrella. They assassinated some of the residents who lived on the collective farms. On 15 July 1997, Tovar called on all those living on the farms at La Pola, El Encanto and La Palizua to attend a meeting at which he communicated that they had 15 days to vacate their properties. Between 1997 and 2000, INCORA revoked the collective land titles because the owners had 'abandoned' the properties. INCORA then transferred them to the names of members of the Bloque Norte of the AUC and Tovar's *testaferros*. In other words, after the paramilitary had massacred and forcibly displaced people, INCORA documented that the owners had abandoned their properties. In 2011, an investigation was opened by the Fiscalia (public prosecutor) against six former employees of INCORA and several employees of INCODER for links to the paramilitary and complicity in forced displacement (Verdad Abierta 2011a). In a similar case, in 1994 Jose Manuel Tirado was threatened and forced to abandon his land in Necocli fearing for his life and his family. He submitted a letter to INCORA explaining the reason why he was leaving the parcel of land he was granted in 1987. He submitted a second letter in 1995 hoping that the institution would prevent him from losing the land, but in vain. In 2007, the land appeared registered in the name of an unknown person who had never even visited the site (Verdad Abierta 2011b).

In addition to INCORA/INCODER, other state institutions have played an important role in legalizing illegally acquired land – a process that again echoes Marx's (1867/1990) findings on the fusion between the propertied class and the political elite as well as the involvement of the state in primitive accumulation in the nineteenth century. For instance, Banco Agrario and the Fund for Financing the Farming and Livestock Sector (Fondo para el Financiamiento del Sector Agropecuario, or FINAGRO) facilitated this process by providing credit and subsidies to the palm agroindustry in spite of accusations made by those displaced and other witnesses pointing to the illegalities surrounding the acquisition of titles over properties where palm oil was cultivated. Public notaries, such as Notarias Unicas in Chigorodo and Carepa in the Department of Chocó, Notarias 5, 18 and 26 in Medellín, and Notaria 8 in Barranquilla, performed the function of legalizing fraudulent transfers of land titles (Verdad Abierta 2011c).

The result of forced displacement, forced abandonment, and dispossession has been a large-scale cross-country counter-agrarian reform through which approximately ten million hectares of land have been transferred into the hands of the upper class over the last 20 years, according to the National Movement of Victims of State-sponsored Crimes (Movimiento Nacional de Víctimas de Crimenes de Estado, or MOVICE)[42] (Comisión Nacional 2009). This counter-agrarian reform has enabled the advent of the palm oil agroindustry as a response to the decline in profitable opportunities in the banana and cotton sectors. For example, in the Urabá region, Department of Antioquia, African palm oil companies have been appropriating the land abandoned by residents fleeing massacres committed by the paramilitary (Donahue 2005). Many Afro-Colombians who had secured land titles in this region have suffered intense attacks by these armed groups. A significant number of the entrepreneurs and promoters of the palm-growing agroindustry, including plantations and processing plants in the Departments of Chocó, Córdoba, North Santander and the region of Magdalena Medio, were/are owned by paramilitary chiefs. Examples include: Inversiones Agropalma which at one point was owned by Jesús Ignacio Roldan Pérez, alias Monoleche, and Palmas S.A. which was owned by Diego Murillo Bejarano, alias Don Berna (Verdad Abierta 2011c). Marx's argument that 'the systematic theft of communal property was of great assistance, alongside the theft of the state domains, in swelling those large farms which were called in the eighteenth century capital farms' (1867/1990: 886) is quite appropriate here with regard to the function that present-day processes of primitive accumulation have served in the establishment of agribusiness farms.

The paramilitary has been known to use the strategy of forced displacement and forced abandonment not only in rural areas with farming properties, but also in commercial and residential areas. For example, when paramilitaries arrived in the town of Calamar, Department of Guaviare, in the early 2000s, they announced that everyone who have been living in Calamar for more than three years had to leave. Subsequently, they repopulated the town and the new residents have been paying rent for housing and commercial establishments to the armed group in control of Calamar, instead of to the legitimate owners (CINEP 2005a).

Para-Political Pacts

The paramilitary has attained widespread dominance over political governance at all levels: presidential,[43] congressional, departmental and municipal. This is expressed mostly in the election of politicians allied with the paramilitary. The paramilitary elite finances the electoral campaigns of such politicians, eliminates their opponents, and often coerces the population into voting for their candidates. By having these political figures in office, the paramilitary influences and even to a large extent controls government decisions in favour of their economic interests and those of the capitalist classes they represent. At the same time, paramilitary friendly politicians facilitate the operation of illegal businesses run by these groups and channel public funds into their hands.

Let us begin by looking at the presence of paramilitary power inside Colombia's congress.[44] Seventy-seven per cent of Congress members elected for the period 2002–6 had links to paramilitary groups (Verdad Abierta 2009a). In 2007, the Supreme Court began investigating numerous connections between Congress members and paramilitary organizations, which gave rise to what became known as the para-politics scandal (*escandalo de parapolitica*). It was discovered that former Vice-President Francisco Santos participated in the creation of the Bloque Central of the AUC and is currently being investigated by the Supreme Court. Former Minister of the Environment Sandra Suárez is also under investigation for links to paramilitary commander Rodrígo Tovar Pupo. Fifty-three senators and members of the Chamber of Representatives, as well as nine governors, have been found guilty by the Supreme Court of links to paramilitary groups. There are 100 cases still in process. Seven of the ten Senate presidents for the period 2002–12 are being prosecuted for ties to the paramilitary. In 2013 former President of the Chamber of Representatives Cesar Perez Garcia was sentenced to 30 years in prison for his part in the Segovia massacre, Department of Antioquia, where in 1988 the paramilitary murdered 43 people (El Tiempo 2013). Former Congress member Alvaro Garcia has been found guilty for orchestrating the Macayepo massacre, in the Department of Sucre, in October 2000, where the paramilitary executed 15 people and caused the displacement of 200 (Verdad Abierta 2012c).

These alliances between members of political institutions and paramilitary organizations were not isolated informal dark deals between corrupt politicians and illegal armed groups, but were rather done in a systematic and often quite formalized fashion and often took the form of 'pacts' (agreements) signed by both sides. For instance, in 2002, former Congress members signed the 'Pacto de Singapur' with paramilitary leader alias El Aleman to ensure that the paramilitary's candidate Julio Ibarguen Mosquera became the governor of Chocó for 2003–7. In 2003, former mayors of the towns of Monterrey, Tauramena, Aguazul, Villanueva,

Sabanalarga and Mani in the Department of Casanare signed the 'Pacto de Casanare' with the paramilitary organization Autodefensas Campesinas de Casanare where they committed to give 50 per cent of their municipal budgets to the paramilitary leader Martin Llanos. In 2002 paramilitary commander Rodrígo Tovar solicited support from politicians in the Department of Cesar to facilitate the election of Mauricio Pimiento as a senator and Jorge Ramirez Ubina as a member of the Chamber of Representatives. In early 2000, the 'Pacto de Chibolo' in the Department of Magdalena was signed, which ensured that the paramilitary friendly Jose Domingo Davila Armenta became the Governor of the Department of Magdalena. Davila was recently sentenced by the Supreme Court to seven years and six months in prison for his links to the paramilitary. According to Alvaro Osorio, chief of the Fiscalia's department responsible for reporting to the Supreme Court,

> First they [paramilitaries] finance the election campaign of their candidate. Once the candidate becomes a governor he would appoint to positions of power individuals who would be favourably oriented towards the AUC and would assign contracts that would feed the [economic] structures of the paramilitary. This way of co-opting the state eventually allows paramilitary power to reach such a point where it can be said that the real power is exercised by them, so the administration of the department begins to function in ways that benefit their interests. (Verdad Abierta 2012c)

By accomplishing such a level of penetration of political institutions, the paramilitary (who are themselves capitalists in terms of the founders, leaders and commanders) ensure that they can control decisions on who would get concessions and contracts from the state and can ensure that their interests would be looked after. As we can see, political power is a means to protecting and advancing economic power and at the same time economic power is a means to securing political power. The fusion of the two means that the state and the capitalist class should always be considered in relation to each other.

The Paramilitary and Other State Institutions: Strategic Symbiosis

The relationship between the paramilitary and two of the key institutions of the state's coercive apparatus, the military and the police, takes many forms, which can be categorized as follows:

1) failure of state forces to intervene in order to protect civilians from paramilitary aggression;

2) tolerance for the establishment of paramilitary bases in proximity to military bases;

3) provision of firearms, other equipment (such as helicopters), uniforms and transportation services by the army to paramilitary units;

4) exchange of captured guerrilla fighters by the paramilitary for weapons and uniforms offered by the military;

5) provision of space for the training of paramilitary fighters in state military bases;

6) exchange of intelligence;

7) sharing of personnel (individuals working at the same time for both the state army or police and the paramilitary);

8) security provision for paramilitary activities by the state army or police; and

9) joint operations (Hristov 2009b).

One example of the latter is the forced displacement that took place along the rivers Salaqui, Cacarica and Truando in the Department of Chocó in the late 1990s as part of two simultaneous operations: Operation Genesis carried out by the Seventeenth Brigade of the Colombian armed forces and Operation Cacarica carried out by the ACCU.[45] Subsequent displacement operations were carried out in 2000 and 2002 by the AUC, clearing the way for the oil palm agroindustry.

Another example, representative of hundreds of cases such as this, took place on 18 February 2000, in El Carmen de Bolívar, Department of Bolívar, where members of the Bloque Norte y Anori of the ACCU under the orders of Emmanuel Ortiz, and with the complicity of troops from Batallion de Fusileros of the First Brigade, executed 46 peasants after torturing them, slashing their throats, and sexually abusing the women. When the paramilitaries were entering the town through the area of El Salado, those who tried to impede their entrance had their throats cut or were hit in the head with a screwdriver. Later, the paramilitaries took out a three-year-old girl from her house and held a knife on her neck in order to make her mother cook for them. According to the testimonies, 'One girl was raped, made to eat cactus, and then left to die choking on her own blood' (CINEP 2005c: 290). Paramilitary penetration of justice system institutions, health care and universities, was also present under the reign of the AUC and did not end with the 2006 demobilization, as the following chapter illustrates.

Social Movements: From Wealth Redistribution to Human Rights to the Revival of Land Struggles

The social movement landscape in Colombia today is very diverse and vibrant – it includes *campesino*, indigenous, women, students and labour union organizations, as well as many human rights advocacy groups and victims' support groups. From the early 1900s to the late 1980s Colombia witnessed the height of rural and urban workers' struggles. These mobilizations were aimed at politico-economic transformations, such as land reform, the right to strike, the right to unionize, improved working conditions and better pay. However, in the period 1990–2010, without dismissing or underestimating the mobilizing capacity, convictions and commitments of these movements for social change, as well as the improvement in strategic organizing among certain movements such as the CRIC, it is possible to observe a lack of advancement with regard to their objectives and accomplishments manifested in the gradual shift in their demands from wealth redistribution to respect for human rights (narrowly defined as the right to live, freedom from harm, etc.). This became particularly pronounced in the 1990s and was accompanied by the emergence of many NGOs dedicated to the protection of human rights. The vision of politico-economic transformation or at least reform became lost or obscured by the more immediate need to protect life. Movements that were previously formed to accomplish improvements in human well-being, such as the labour or women's movements,[46] were now being so frequently attacked and partially exterminated that their activism became overwhelmingly focused on denouncing the deaths and abuses against their members instead of advancing any other cause. Pleas to protect human rights became the end goal. Many of the struggles for land that we see today are led by people who have been forcibly displaced and hope to recover what was stolen from them to begin with. A present-day example of this is the loose associations of families of those victimized by state and paramilitary violence. The demand of past peasant struggles for an 'end to the *latifundio*' was not heard among non-armed social movements in the 20-year period from 1990 to 2010. These setbacks, in my opinion, have been caused by the increased strength of the state's coercive apparatus and paramilitary offensives.

However, 2012 and 2013 have been characterized by a reawakening of massive popular mobilizations and the clear articulation of new social movements with radical demands, a large part of which are centred on the agrarian question. One newly formed movement is the Movimiento Politico Marcha Patriotica founded in April 2012 and comprising various organizations from the Left, including the National Association of Areas of Peasants Reserves (Asociación Nacional de Zonas de Reservas Campesinas, or ANZORC), the Peasant Association of Cimitarra River Valley (Asociación Campesina del Valle del Río Cimitarra, or ACVC), the National Coordination of Agrarian and Popular Organizations (Coordinación Nacional de Organizaciones Agrarias y Populares, or CONAP), the Federation of Agricultural Workers and Small-Scale Farmers (Federación Nacional Sindical Unitaria Agropecuaria, or FENSUAGRO), the PCC, and the Colombian Communist Youth (Juventud Comunista Colombiana, or JUCO). Among Marcha Patriotica's leaders are Piedad Cordoba, Gloria Cuartas, Gloria Ines Ramirez and Jaime Caycedo Turriago. The movement unifies the struggles of the social sectors most disadvantaged by capitalism. Its objectives revolve around agrarian reform, reparations for the victims of the armed conflict, and popular sovereignty. In 2013 former senator and leader of Marcha Patriotica, Piedad Cordoba, announced that since the movement came into being two years earlier, 29 members had been assassinated, three had disappeared, and trials were underway against 200 members. All of this has led the leaders to consider the possible dissolution of the movement in order to protect the lives of its members (NODAL 2014).

The second half of 2013 saw a wave of agrarian uprisings which were first expressed on a local level but eventually swept most of the country. On 11 June 2013 in the region of Catatumbo, Department of Norte de Santander, a strong popular mobilization for economic, political, cultural and environmental rights led by the Peasant Association of Catatumbo (Asociación Campesina del Catatumbo, or ASCAMCAT) and the Inter-sector Committee of Catatumbo (Comite Intersectorial del Catatumbo, or CISCA) took place. There were several factors prompting the uprisings. Catatumbo is a region rich in oil, carbon, uranium and gold. It is dominated by coca production, agroindustries and mining. In 2012 the government declared it a strategic mining zone. Fifty-seven mining titles have been given to local and foreign mining companies. At the same time, much of the population in this region lacks access to electricity, running water, water-treatment plants, health-care facilities, paved roads and reliable means of transportation. People drink water that has been polluted with human faeces. Coca leaf production is the only viable source of income for the rural poor. In 2012 the government initiated manual coca plant eradication operations while completely disregarding the 'Sustainable Development Plan for the Creation of Areas of Peasant Reserves' (Plan de Desarollo Sostenible para Construir Zonas de Reservas Campesinas) formulated by local small-scale farmers as a practical alternative that would allow them to abandon the cultivation of illegal crops.[47] The creation of Areas of Peasants Reserves (Zonas de Reservas Campesinas, or ZRC) was authorized by Law 160 of 1994. The ZRCs represent a concrete possibility for addressing the extreme concentration of landownership by having the state provide landless rural residents access to land which would be exempted from mining and agribusiness activities and at the same time would help to promote food security and environmental sustainability. To be approved for the creation of a ZRC, local residents have to provide a development plan that outlines how they would sustain themselves economically off the land. If for instance, the ZRC has an area of 100,000 hectares, the state would have to ensure that each owner within a ZRC would have no more than 200 hectares. It would also be responsible for providing sanitation, electricity, health-care facilities and other infrastructure. Only six ZRCs exist in the entire country (Celis 2013).[48]

The rural population in Norte de Santander has been demanding the creation of a ZRC for the past 20 years to no avail. Thus, the mounting discontent over the lack of basic infrastructure, and the failure to establish a ZRC, combined with coca eradication operations without the consent of the local population, burst into protests with over 25,000 rural residents from eight municipalities setting up highway blockades surrounding the towns of Tibu, Ocaña and others. Talks were scheduled to take place on 3 July 2013 between demonstrators and government officials including the Agricultural Minister Francisco Estupiñán, Norte de Santander Governor Edgar Díaz, head of INCODER Miriam Villegas, and the mayors of ten local municipalities in the presence of Army General Rodolfo Tamayo and Police General Rodolfo Palomino. On 15 June, a heavy confrontation took place for seven hours between the Anti-Disturbances Mobile Squadron (Escuadron Movil Antidisturbios, or ESMAD) and protesters in an attempt by the former to clear the highways. At the same time, state armed forces attacked 400 peasants with tear gas near the municipality of Hacari, while robbing them and causing injuries. By 22 June, ESMAD forces in Ocaña were joined by 600 police and soldiers. Two demonstrators were killed, 50 were seriously wounded (among them women and children), and hundreds were arrested. The UN High Commission for Human Rights in Colombia announced that ESMAD had exercised excessive use of force against the protesters. President Santos suspended talks and rejected the proposed ZRC. ASCAMCAT leaders then solicited a meeting with Vice-President Angelino Garzon which was eventually obstructed by Santos. While these confrontations were taking place in Catatumbo, representatives of peasant movements from the Departments of Guaviare, Meta, Cauca, Nariño, Caqueta, Quindío, Putumayo, Sucre and Antioquia met in Bogotá to plan a national agrarian strike in solidarity with the demands of the people of Catatumbo, especially the ZRC under question (Duque 2013).

In August 2013 the Roundtable for National Agrarian Dialogue (Mesa de Interlocución Agraria Nacional, or MIA) comprising various agrarian organizations at the municipal, departmental and national level, called for the National Popular Agrarian Strike beginning on 19 August. The strike or shutdown, which lasted 21 days, became an event of historic significance since it united agricultural workers from various sectors, small-scale farmers and landless people, along with the support of students and other social movements, who marched in solidarity. The argument that the Left criticizes capitalism but offers no solution has been heard over and over again from academics, politicians and others convinced that there is no alternative to capitalism. The peasant movement of Colombia showed the whole world that this is not true by presenting a long set of concrete proposals as to how processes of impoverishment and marginalization can be reversed. These illustrate the crucial role that the state must play. Among the key demands that the MIA sees as obligations that the government must fulfil are:[49]

1) establish fixed prices on food crops grown by small-scale farmers sold at local markets to guarantee a basic level of income for the producers and access to food supply for the consumers;

2) create a national compensation fund that would assist small-scale farmers whenever the cost of production is higher than the income from sales;

3) establish a guaranteed percentage of the produce to be purchased by the state from small and medium-scale producers and to be distributed to city markets;

4) reduce gas prices and toll fees;

5) establish price controls for fertilizers, pesticides and other farm supplies;

6) recognize that small and medium-scale farming is the way towards achieving food security and environmental sustainability and implement policies that support such producers and promote clean and environmentally friendly agriculture;

7) discontinue current anti-drug policies and establish new policies for the gradual substitution of coca, marijuana and poppy cultivation;

8) formulate legislation that protects traditional and ancestral seeds;

9) stop the importation of agricultural food products such as potatoes, milk, rice, coffee and cocoa;

10) stop and revise the free trade agreements currently in effect with the US, Canada, the European Union and other countries;

11) pardon the debts accumulated by small and medium-scale farmers;

12) create subsidies for small and medium-scale farmers through the public bank;

13) establish crops insurance that would protect farmers in cases of natural disasters or epidemics. This insurance should be provided by the state without the involvement of private financial institutions;

14) fulfil the obligations under Law 160 from 1994 by purchasing sufficient land of good quality and allocating/granting land titles to landless rural residents accompanied by investment in technology, technical assistance, and marketing by the state to ensure the viability of small and medium-scale farmers;

15) apply immediately Decree 1277 of 2013 in the granting of collective land-titles by INCODER to indigenous and Afro-Colombian communities and recognize the autonomy of these communities over the granted territories;

16) constrain the foreign ownership of land. Cancel or revert transactions or operations that have allowed the monopolization of land by foreign companies and use the recovered land as part of the Land Fund for the allocation and granting of plots to landless rural residents;

17) within one year complete the process of redistributing idle large-scale territories to landless rural residents;

18) demarcate the ZRCs in cases where residents have already presented their proposals and all required documentation for their constitution within one year of their solicitation;

19) speed the processing of solicitations for expansion of *resguardos* and collectively owned territories by Afro-Colombian communities;

20) suspend the concession of mining titles and revert concessions already made to big companies. Formulate a new mining policy with the consultation of artisanal/small-scale miners that would take into consideration the well-being of surrounding communities;

21) create a new Mining and Natural Resource Exploration Code that redefines the maximum terms for mining exploration in any given area, ensures mechanisms to mitigate the negative effects of such activities and promotes artisanal and medium-scale mining;

22) democratize the National Federation of Coffee Producers;

23) guarantee access to primary, secondary and technical education for all rural inhabitants;

24) abolish Law 100 and formulate new health-care legislation;

25) discontinue the privatization of public service provision in sanitation and plumbing and ensure that these are operated by municipalities; and

26) allocate resources for housing construction (Agencia Prensa Rural 2013).

The unarmed protests in different parts of the country were met by ESMAD with the use of stun grenades, rubber bullets, tear gas, water tanks and bullets. Over the course of the 21-day strike, according to MIA there were 660 cases of human rights violations including 52 cases of harassment and threats, 262 arbitrary detentions, 485 injuries (21 of them by firearms) and four people disappeared (Bonilla 2013). In the case of the Catatumbo uprising as well as the National Popular Agrarian Strike, Santos' administration accused the protests of being organized and infiltrated by the guerrilla, a claim that has been conveniently used for decades to divert attention away from the protesters' demands, to criminalize them, and to justify the state's use of excessive force.

4

Eight Years After the Demobilization of the AUC: The Predatory State-Capital Alliance of Domination (2006–2014)

Even after the demobilization of the AUC, paramilitarism has continued to be an essential instrument for capital accumulation by displacing *campesinos* to free up land for extractive industries and agribusinesses as well as by repressing labour unions and other forms of social activism. This chapter reveals the continuation of paramilitarism in Colombia after 2006 by first documenting the activities of illegal armed groups[1] in the 2006–14 period and interrogating the notion of BACRIM (an acronym for *bandas criminales* or criminal gangs) as an ideological construct used by the state and media discourses to create the illusion of a rupture between a paramilitary past and a 'post-paramilitary' present. This is followed by an examination of the composition, activities, and links to state institutions of illegal non-guerrilla armed organizations, exposing their paramilitary nature. I then discuss the inefficiency and inadequacy of the legislative tools and processes that have been introduced to bring about justice and reparations for the victims of paramilitary violence in order to illustrate how the resulting impunity is conducive to the continuation of the military and economic structures of paramilitarism.

The following is only a very brief background on the demobilization of the AUC. On 23 December 2002, President Uribe authorized a law that enabled negotiations to be carried out with any illegal armed groups, including the guerrilla (CINEP 2002). On 15 July 2004 an agreement was reached between the government and the AUC, known as the Accord of Santa Fe de Ralito (Acuerdo de Santa Fe de Ralito). The Accord stipulated that the organization begin the gradual demobilization of around 49 paramilitary blocks spread over 28 out of the 32 departments of the country, to be completed by the end of 2005 (CINEP 2002). Uribe's administration did its best to accommodate the needs of these groups with the help of several legislative measures, such as Decree 128 of 2003 oriented towards pardoning their crimes. Giving up illegally acquired land, performing community service, or paying a fine and leaving the country were among the alternative 'punishments' (Gutiérrez 2005). The Justice and Peace Law (Ley de Justicia y Paz) was approved by Congress on 21 June 2005 and by the Constitutional Court in May 2006 after the latter made some modifications to the original version. It offers reduced prison sentences, financial benefits (such as government stipends), and readjustment training for those who demobilize. Prison sentences are limited to a maximum of eight years (Gutiérrez 2005). By March 2006, the media announced that 31,671 had given up their arms which meant that all the paramilitary blocks belonging to the AUC had demobilized (Restrepo and Franco 2007). There were, however, several serious reasons to question the authenticity of the demobilization process:

1) paramilitary groups continued to engage in human rights violations during the ceasefire;
2) a considerable number of those who participated in the demobilization ceremonies were in reality not paramilitary combatants but hired *sicarios* who pretended to be AUC fighters;
3) the quantity of weapons turned over was much less than what the paramilitary were known to possess;
4) the subsequent activities of those who laid down their weapons provided them with access to the means of violence; and
5) the legal and political measures employed to regulate the demobilization process and its aftermath were highly conducive to impunity (Hristov 2009b).

The Invention of Bacrim and the Invisibilization of Paramilitarism

Human rights organizations, NGOs, community organizations, social movements, some research centres, and political figures have been documenting cases of violations against civilians carried out by illegal armed groups after the so-called demobilization in 2006. Contrary to claims advanced by the Colombian state and media, based on my research I argue that what should have been a 'post-paramilitary' era has in fact been characterized by a continuation and in some cases strengthening of the military and economic structures of paramilitarism. In fact, since the demobilization was completed, paramilitary groups have been reorganized, recomposed and 'cleansed' (meaning they have ridden themselves of allied groups and individuals who are no longer useful to the paramilitary system for a variety of reasons), making the restructured networks more efficient. Among the groups that have appeared since 2006 are: Las Águilas Negras (in various regions), Los de Magdalena Medio (in the region of Magdalena Medio), Autodefensas Campesinas del Pacifico (Department of Valle del Cauca), Autodefensas Campesinas Nueva Generación (Department of Nariño), Autodefensas Unidas de Antioquia (Department of Antioquia), Autodefensas Gaitanistas de Colombia (Departments of Urabá, Córdoba, Medellín, Cauca), Los Rastrojos (Northern Department of Valle, as well as Departments of Quindío, Chocó, Nariño, and Cauca), and Ejército Revolucionario Popular Anti-Subversivo de Colombia-ERPAC (the Eastern Plains, also Departments of Meta, Casanare, Guaviare, Arauca and Vichada), Los Paisas (Department of Magdalena), Los Machos (Department of Valle del Cauca), Renacer (Department of Chocó), Bloque Conquistadores (Valle del Cauca), Muerte a Sindicalistas (Barranquilla), Los Macacos (Departments of Meta, Vichada, and Casanare), Mano Negra (Barrancabermeja) and Los Urabeños. These organizations are not completely new, since they include former AUC combatants and mid-level commanders who have rearmed, members of old groups that officially remained active during the peace process and never demobilized (such as the Bloque Cacique Pipinta), new recruits, criminal gangs, and police and military officers, as well as even mayors and governors (Hristov 2009a).

Human rights agencies and social movements have reported that these groups engage in extortion, smuggling of resources, drug-trafficking, forced displacement, selective assassinations, social cleansing and various forms of attacks and intimidation against social movements, activists and Leftist students and academics (Hristov 2009a). For instance, during the three months after the completion of the demobilization process, paramilitary groups such as the Águilas Negras and the Autodefensas Campesinas Nueva Generación sent out numerous death threats to human rights defenders, the National Indigenous Organization of

Colombia (Organización Nacional Indígena de Colombia, or ONIC) and the Confederation of Colombian Workers (Central Unitaria de Trabajadores, or CUT). One of them stated 'we are making clear to you the AUC were our base and have completed one phase of their service to the people of Colombia, and once the processes of demobilization have been completed, we are now the present and future of the Colombian state for years to come ... we are present in 21 rural and urban areas ... and we operate in a variety of forms' (AI 2006b). On 6 August 2011 in the rural area of the municipality of Tierralta, Department of Córdoba, 40 armed men arrived in the neighbourhood of Paila and massacred six young men. This was the fourth massacre in 2011 carried out by the 'new' illegal armed groups. Manuel Wilches, who lost his son and two sons-in-law, said 'This assumption that it [the violence] is gangs against gangs is false. We are peasants, and they killed my young ones' (Semana 2011d). Some of these armed groups have achieved a high degree of territorial control in a way that very much resembles the AUC regional domination. For instance, on 4 January 2012, Los Urabeños paralyzed 16 municipalities in the Department of Cordoba by ordering citizens to suspend any commercial or transportation activities until further notice. From public transport systems to milk pasteurization plants, economic activities came to a halt. Mayor Gabriel Alberto Calle of Monte Libano, Department of Cordoba, recognized that the shutdown would end only when Los Urabeños authorized it (Semana 2012a).

As noted above, the 'new' illegal armed groups have been labelled by the Colombian state with the acronym 'BACRIM' – *bandas criminales* or criminal gangs presumed to be at the service of drug-trafficking. According to a statement by the Colombian Ministry of Defence in March 2006, 'these delinquent gangs are in many cases hired by FARC and in other cases are a product of the recruitment carried out by drug-traffickers who seek to form their private security groups' (Restrepo and Franco 2007: 66). Moreover, in March 2006 the government's High Commissioner for Peace, Luis Carlos Restrepo, claimed that these groups 'are not self-defense [that is, paramilitary]. What we have in various areas of the country are very small emerging criminal organizations which are managing illegal crops ... these organizations are completely dedicated to drug-trafficking and on many occasions also combine this with extortion. We cannot call them self-defense' (Restrepo and Franco 2007: 66). During his presidential speech at the National Police Commanders Summit on 30 January 2007, former President Uribe ordered that paramilitarism should no longer be spoken of. Subsequently, key figures from the Colombian state military, such as Colonel Ricardo Restrepo Londono (El Tiempo 2009) – one of the official experts on BACRIM – as well as members from the National Police, political analysts and mainstream media, all declared that existing armed groups were small in size and concentrated on the administration of illegal businesses rather than counter-insurgency. Even the General Secretary of the OAS stated at one point: 'the armed factions that emerged after the demobilization ... have a criminal profile linked to drug-trafficking. There has not been any evidence of counter-insurgency actions that would link these organizations to the concept of paramilitary' (Gaviria et al. 2008: 39). To emphasize that paramilitarism is history, the Colombian government's Department of Social Prosperity (Departamento para la Prosperidad Social) created the Historic Memory Centre dedicated to compiling testimonies of those who have been victims of human rights violations by any armed groups, including the paramilitary. Plans are underway to make exhibits of images and texts at the National Museum illustrating the victims' experiences.

What do the media have to say about all this? It is true that Colombian media (especially those claiming to be politically neutral and analytically oriented, such as *Revista Semana*) have made public some of the confessions of detained and extradited paramilitary commanders that illustrate past connections between the AUC and politicians or the state military. However, the key word here is 'past'. While the public's attention is consumed by shocking revelations about past deeds of the para-state partnership, which most often become known in the form of scandals (such as *parapolitica* [para-politics], *falsos positivos* [false positives], and *chuzadas* [illegal interceptions/surveillance]), the ongoing present re-consolidation of paramilitary groups and their terror strategies continues unchallenged and unacknowledged. The word 'paramilitary' is used only in the context of events that took place prior to the demobilization. There is a similar tendency even in some human rights organizations. For instance, in a March 2010 report by the UN High Commission on Human Rights, under the section 'Violations of International Humanitarian Law', only two headings appeared. One was 'Violations by State Forces' and the other 'Violations by the Guerrilla'. There was no mention of 'paramilitary'. Surprisingly, a similar approach has been taken by some other independent media outlets and research centres. The language of the Colombian state's ideology, which insinuates a rupture between the past and present and erases the paramilitary from the country's present violent landscape, has indeed been so powerful and prevalent that it has had an effect even on some of the few media dedicated to denouncing human rights violations by the paramilitary and the state. Only recently have some journalists begun to use terms such as 'neo-paramilitaries' or 'paramilitary's heirs' to question the presumed distinction between past paramilitary and present criminal armed groups.

As we can see, according to official state discourses, media and some human rights agencies, paramilitary groups no longer exist, and the term *paramilitary* has been definitively replaced with the label *criminal gangs* to refer to any non-guerrilla armed group. With the invention of BACRIM, the government has managed to virtually erase the presence of the paramilitary from the picture of the armed conflict. The Colombian state, however, is not alone in this fabrication of the end of paramilitarism. Academic writings can reinforce the myth through fragmented de-historicized characterizations of the paramilitary. Such writings were discussed in Chapter 2.

The disappearance or invisibilization of paramilitarism, as a discursive strategy used by the state and some scholars, bears some similarity to the abandonment of the subject of political violence during the post-war reconstruction and peace-building period in Central America from the 1990s up to the present. In this context, broadly speaking, parallels between Central America and Colombia can be drawn in terms of the continuity in existing forms of repression as well as the proliferation of some new ones, the shift in the dominant security discourse particularly with regard to the justifications offered for repressive forms of social control, and the ultimate purpose that all of these have served. Pearce (1998) argues that in Central America, beginning in the early 1990s, Cold War ideology around the Communist threat was gradually replaced by issues around drug-trafficking and organized crime. After the peace accords in El Salvador (1992) and Guatemala (1996), the new sources of disorder were no longer the guerrillas but rather criminals, drug mafias and youth gangs. These became the official target of the state's coercive apparatus. In reality however, political violence carried out by parainstitutional armed bodies linked to the state military has persisted in both countries and, thus, the process of democratization has been accompanied by the systemic violation of human rights particularly among peasant movements and human rights defenders. Moreover, in recent years, new security initiatives, such as CARSI (discussed in Chapter 1), have been put in place, the aim of which is to enable partnerships between the state military and police on one hand, and private armed groups on the other, under the slogan 'public-private partnerships'. The ultimate purpose such security projects serve remains essentially the same as it was at the beginning of the 1990s – to advance economic liberalization and enable the penetration of national economies by global capital. As Pearce (1998) remarks, the debates around institutional, electoral and political reform that were part of the post-Cold War reconstruction were linked to discussions around economic privatization and structural adjustment. The state and the local elites allowed the political participation of previously excluded groups in return for their acceptance of capitalist modernization and renouncement of agrarian reform and other wealth redistribution demands.

In sum, the invisibilization of the subject of political violence in Central America from mainstream media, state discourses and large parts of the academic literature has served the purpose of obscuring the continuation of the political character of a considerable portion of the violence enacted by state and parainstitutional actors against those who contest the neoliberal model, while allowing for the introduction of a novel justification for state-sanctioned violence (fighting organized crime and drug-trafficking). In a somewhat similar fashion, in Colombia the introduction of BACRIM plays a twofold purpose: enabling the disappearance of the paramilitary and its links to the state (that is, political violence) as well as providing a pretext for the continuing expansion and enhancement of the state's coercive apparatus both quantitatively and qualitatively, which in turn secures the conditions for further capital accumulation.[2]

Paramilitary Groups in a 'Post-Paramilitary' Era: Dispossession and Repression

This section will show why the illegal armed groups labelled BACRIM are paramilitary in nature and not merely criminal gangs.

Leaders and Members of Illegal Armed Groups: The AUC's Successors

In many cases, groups that are deemed by the government to be merely criminal gangs, such as Los Rastrojos and Águilas Negras, are led by and composed of former AUC fighters. The Águilas Negras is an organization that has been terrorizing the civilian population in different parts of the country since the AUC demobilized. It was founded inside the official demilitarized zone of Santa Fe de Ralito during peace talks with the government in 2005 by Carlos Mario Jiménez, alias Macaco, an AUC commander responsible for numerous massacres (Semana 2009c). The Águilas Negras comprises former fighters of the Bloque Central Bolívar and three of its leaders in 2007 were former AUC members. Leech quotes one of the leaders of the Afro-Colombian movement on the Pacific Coast (Proceso de Comunidades Negras, or PCN), as saying 'Only the name is different. They are the same people. The top commanders have gone; the new commanders are those who previously were second and third-level commanders' (2009: 31). According to him, new paramilitaries collude with the army and the local political establishment just as before.

In a similar fashion, soon after the demobilization of Rodrígo Tovar, alias Jorge 40, and his men of the Bloque Norte, another armed group, Los 40, was born. Some of the fighters who used to belong to the paramilitary organization led by Diego Fernando Murillo, alias Don Berna, formed the group Los Paisas. Former combatants who used to work for paramilitary chief Hernán Giraldo Serna created Los Nevados. All of these organizations are active on the Caribbean Coast (Semana 2007b). It has been estimated that over 50 per cent of the chiefs of BACRIM groups are former AUC members (Semana 2011c).

There is evidence showing that several important former paramilitary leaders currently in jail continue to coordinate illegal activities. In August 2007, various recordings of conversations between jailed paramilitaries at the High Security Prison of Itagui and their men at large were made public. One of these was a discussion between Elkin Triana and another member of his organization outside of jail, known as El Mono, about the sale of narcotics. Another conversation took place between paramilitary member alias Goyo in the Itagui jail and a member of his paramilitary organization who called him on his cellular phone to give him a report on the re-armament of his men and the taxation of residents of various neighbourhoods in Medellín by members of this group (Semana 2007a).

Illegal Armed Groups and the State: Condemnation or Cooperation?

While the Colombian government claims to have captured more than 13,000 and killed 1,300 BACRIM members between 2006 and 2011 (Semana 2012b), the profound penetration of major state institutions by paramilitary power, illustrated in the preceding chapter, remains in place today. State involvement through complicity, tolerance, collaboration and direct participation in paramilitary activities has not ceased. Colombian police and military personnel collaborate and often directly participate in the operations of the so-called BACRIM groups as they did with the AUC in the past. For example, the group Los 40 was formed when a former police officer, Miguel Villarreal Archila, alias Salomon, brought together demobilized paramilitary fighters as well as members of the state police and armed forces. On 31 August 2007, when 50 members of Los 40 were arrested, it was discovered that 18 of them were active police officers (Semana 2007b).

In 2011, it was found that the leader of Los Rastrojos, alias Sebastian, had a long list of active police, military and DAS officials on a payroll. In a period of only four months, these officials had received payments from Los Rastrojos that totalled approximately $360,000 (Semana 2011d). Of the 24 Los Rastrojos members captured in Chocó in May 2011, there were seven active police officers, two navy sub-officers, a CTI investigator, a secretary from the municipal court and a municipal council. In 2011, the police dismissed 300 officers for connections to BACRIM groups and DAS dismissed 30 of its agents for the same reason. In the same year, 350 army members and 12 prosecutors were under investigation for links to illegal armed groups. In 2010, the CTI captured an army colonel and a lawyer employed in the Fiscalia who for two years stole army weapons and other equipment and sold them to BACRIM (Semana 2011c).

Politicians' alliances with paramilitary organizations have also been a constant feature of the post-demobilization era. For example, the governor of Guaviare, Óscar López, elected in 2007, had a close relationship with paramilitary commanders Vicente Castaño and Pedro Oliverio Guerrero Castillo, alias Cuchillo (who died in 2010). The armed bodies under the command of these two paramilitary chiefs helped Lopez obtain thousands of acres of land in the department of Casanare and win the 2007 elections, despite scandalous evidence against him (Semana 2009a). In another example, the political project of paramilitary groups in the Department of Chocó continued after 2006 as part of the 'Pact of Singapur' (*Pacto de Singapur*) where the paramilitary leader alias El Aleman, by funding the electoral campaign of Julio Ibarguen Mosquera and administering repression against his competitors, ensured that this paramilitary friendly politician became governor of the department for 2003–7. The same paramilitary groups later supported Patrocinio Sánchez Montes de Oca, who became the subsequent governor of Chocó for 2008–10 (Verdad Abierta 2012c). Between 2003 and 2007, Matias Oliveros del Villar, former mayor of El Banco, Department of Magdalena, along with his wife, used to oblige all employees of the municipality to give 40 per cent of their monthly salary to paramilitary groups in the area (Verdad Abierta 2012b).

Since the para-politics scandals which first erupted in 2007, some of the smaller political parties such as Colombia Viva, Colombia Democratica and Convergencia Ciudadana, which had large portions of their membership investigated for links to paramilitary groups, ceased to exist. But the colleagues, families, relatives and friends of those politicians under investigation formed parties with new names to participate in the 2010 legislative elections. For instance, former members of the three parties mentioned above created the new party Alianza Democratica Nacional (ADN). In fact, Juan Carlos Martínez, a former senator from Convergencia Ciudadana, participated in the organization of the ADN even though he was in prison. Similarly, former members of Convergencia Ciudadana created the Partido de Integración Nacional (PIN) (Salas 2010). This practice of recycling para-politicians by allowing them to be replaced by family members and friends has not been limited only to the smaller parties. For instance, Marta Curi de Montes, a member of the Conservative Party who aspired to become a senator in 2010, is the daughter of Nicolas Curi, who had been sentenced to four years in prison for involvement with paramilitary groups, and is also the wife of William Montes who was implicated in the para-politics scandal but was absolved. In another example, Arleth Casado, wife of the former Liberal congress member Juan Manuel Lopez who had been sentenced to six years in prison, became a candidate for the Senate in 2010 (Semana 2010a).

Another example of the continued presence of paramilitary power inside state institutions can be detected in the practices of INCODER. As demonstrated in Chapter 3, through various mechanisms INCODER has been legalizing land theft by issuing land titles to paramilitary chiefs for parcels they have acquired illegally, while revoking the land titles of the victims of paramilitary terror. Since the demobilization of the AUC, in the absence of any viable state solutions to the problems of displaced people, many of the victims have joined together to demand land restitution and have often attempted to recover their lands themselves. INCODER has been one of the key agents in creating and re-enforcing obstacles to such initiatives by playing an important role in the war for land during the post-demobilization era. Several patterns can be observed in the manner in which INCODER has been supportive of the paramilitary land-grabs since 2006.

In the first instance, families who were forced to flee their homes in the late 1990s and early 2000s eventually lost their land titles since, according to INCODER, they were away from their land for more than five years, which in itself constitutes 'abandonment' of the property.[3] Thus, their titles were revoked without any investigation of the circumstances under which 'abandonment' took place. This was the case with many of those displaced by Rodrígo Tovar in the region of Magdalena Medio (Verdad Abierta 2009b). In recent years, especially since 2006, as many of these victims have returned to recover their properties, INCODER in turn claims to have given these titles to other families in need, but in reality the property appears in the names of fictitious persons and companies, third parties (relatives of a paramilitary owner), or *testaferros*. The end result is that the properties remain in the hands of those who had illegally appropriated them.

The second pattern has been one where displaced people have returned to reclaim their lands and still hold titles. Most of these attempts, however, have not only been unsuccessful but have also had high costs, as community leaders have been threatened and some killed. Let us look at the example of the Afro-Colombian communities in the area of Curvarado and around the river Río Sucio, Department of Chocó. Beginning in 1997, due to combat between the state army and the guerrilla, and the subsequent arrival of paramilitary forces, at least 15,000 residents left their homes and the land on which they used to plant corn, yucca and plantains. Ten years later, some of these people took the brave decision to return. The surprise awaiting them was that all their land, including that on which their homes had once stood, had been cultivated with African palm oil. The present plantation workers live in nearby shacks made out of wood and plastic without any basic sanitation, and use the former community's cemetery as an outdoor toilet. This is not an isolated case. In the Department of Chocó, 29,000 hectares of land to which Afro-Colombian communities hold collective titles is occupied by agribusinesses; 7,000 of it is cultivated with African palm oil. Most of those who returned with some hope have given up and left. But a few of them built small wooden shacks in the nearby humanitarian area established with the help of NGOs, and are currently waiting for the government to intervene so they can recover the land that legally belongs to them and which is part of their history and culture. Yet neither the Inter-American Human Rights Court, nor state forces, nor INCODER, have been able to reverse the harm already done. On the contrary, leaders of the displaced Afro-Colombian communities, such as Orlando Valencia and Walberto Hoyos, have been assassinated. As Ramon Salinas, a 65-year-old man who has been through this whole ordeal, explained: 'I have already endured everything – they killed my brother ... took away my land, destroyed my town, threatened me, displaced me, and today I am starving' (Semana 2009d).

A third kind of pattern in the continuation of dispossession in the post-demobilization era has been one in which small-scale farmers forced to abandon their land and now living in extreme poverty and hunger on the periphery of large cities eventually turn to INCODER asking to be granted a small parcel of land on which to plant subsistence crops in order to meet the basic food requirements of their families. INCODER has often turned down such requests, stating that they are ineligible for such land grants because a registered property already appears under their name (that is, they are not propertyless). In reality, this is the very property that the victim was forced to abandon and is now already owned and occupied by a third party (Agencia de Prensa 2010). In other cases, when *campesinos* return to their lands with valid titles, INCODER informs them that they owe large amounts on a government loan, since they have not been making payments for a number of years. This is of course absurd, because once the residents were forced to abandon their land, they could no longer engage in farming activities in order to afford the payments. Nonetheless, if they wish to retain their titles INCODER expects them to begin paying the unsustainable debt that has been incurred in the intervening years. Facing financial ruin, their only option is to sell their land to the first bidder they get, a sale pre-arranged between INCODER and the investor who has been waiting to take hold of the property.

Frequently, many of the small-sized properties that were amassed by paramilitary commanders and their families and allies through forced displacement were subsequently combined into larger properties (since displacements of numerous small-scale farmers took place on neighbouring plots). The newly formed large estate was then registered under the name of a mock company (registered on paper but without any real operations). The company, owned by the paramilitary, in turn sold the estate at a high price to a reputable enterprise. The buyer then often purchased in good faith, without necessarily being aware of the dark origins of the deal, and set up cattle-ranches, reforestation projects, or cash-crop plantations on the land. This has happened repeatedly in different parts of Colombia. The entire process involves various illegalities and this is precisely where INCODER comes in. One specific example to illustrate the point here is the fertile region Montes de María, located between the Department of Bolívar and Department of Sucre. In Montes de María, 3,128 small-scale farms whose owners were displaced were declared by the government as 'under protective measures' since 2004, after the victims officially declared their wish to recover the lands they were forced to abandon. The meaning of this status of 'protective measures' is that while investigations are going on with regard to the status of the properties, the latter cannot be sold without authorization both from the previous owner (the victim) and from the Local Committee for Protection of the Displaced. However, since 2008 INCODER has allowed such transactions to occur without the authorization of all members of the committee or alternatively has assigned a property a new registration number so that it will not appear among the list of 'protected' properties (Semana 2011a).

In October 2010, INCODER launched the 'Shock Plan for Restitution and Formalization' the objective of which was to re-establish the property rights of 459 families who had been forcibly displaced from their land. Although, under the orders of the Supreme Court of Justice, INCODER issued administrative acts to re-establish the property rights of 32 families, none of these victims have yet recovered their properties due to INCODER's lack of initiative in ensuring that the properties would be vacated by their current owners and made available to the victims (Castro 2012). In sum, as the Colombian Minister of Agriculture Juan Camilo Restrepo has put it, 'INCODER is damaged and needs a total re-engineering' (Semana 2010a).

The continued involvement of INCODER in dispossession (i.e., present-day processes of primitive accumulation) is a clear illustration of Marx's insistence that a productive system is made up of its specific social determinations – specific social relations, modes of property and domination, legal and political forms. As Wood (1981) helps us understand, contrary to the strict 'base-superstructure' interpretation of Marx's theory, the productive base itself exists in the shape of social, juridical and political forms – in particular forms of property and domination. INCODER's activities demonstrate precisely this point – as a state institution it represents the legal counterpart of the system of accumulation present in Colombia. As Marx and Engels put it: 'The conditions under which definite productive forces can be applied are the conditions of the rule of a definite class of society, whose social power deriving from its property, has its practical-idealistic expression in each case in the form of the state' (1846/1968: 60). In the Colombian context, INCODER is one component of the state that secures the conditions for the rule and continuous enrichment of the landowning class.

As mentioned in Chapter 3, the university in Colombia is another institution not immune to the influence and control of paramilitary groups. Once again, this continues to be the case after 2006 as the following examples illustrate. In 2007, Jaime Alberto Camacho Pico, the Dean of the Industrial University of Santander (Universidad Industrial de Santander), collaborated with a member of the paramilitary to compose a list of students, professors and administrative personnel who were thought to be 'following the steps of the Left'. The Dean and some of his colleagues from the university administration attempted to pressure journalism students from the Communication and Television Department of the University to conduct video-recordings of student protests and meetings (Semana 2009c). In 2009, the names of 15 people from the University of Antioquia, including student activists, appeared on a list signed by the Bloque Antioqueño paramilitary group that was circulated among the university community and threatened these individuals with death. In addition, Alfonso Monsalve Solorzano, the Vice-President of Research at the University of Antioquia for 2006–9, when he was the Dean of the Faculty of Human Sciences, used to be paid $2,000 a month by the former director of DAS, Jose Narvaez, for providing information to DAS about suspicious students, professors, researchers and PhD candidates, many of whom were later assassinated by the paramilitary (CJL 2010).

The embezzlement of public funds and the misuse of public facilities, such as hospitals, by government officials who channelled such resources to paramilitary groups, have also continued in the post-demobilization era. Between 2003 and 2007 Angel Maya Daza, former manager of the state hospital Rosario Pumarejo Lopez in Valledupar, Department of Cesar, had given paramilitary groups access to the financial and administrative apparatus of the hospital. He and other colleagues administered payments to the fictitious companies Dismed Ltda. and Ingemedical which belonged to the paramilitary Enrique Guevara Cantillo of the Bloque Norte of the AUC. The manager also used the hospital's ambulances to transport paramilitaries as well as their weapons and equipment (Verdad Abierta

2012a). The presence of paramilitary power in the health-care sector in the Department of Bolivar is also notorious.[4]

Capitalist Predators

According to a 2010 report by Human Rights Watch:

> The successor groups are committing widespread and serious abuses, including massacres, killings, forced disappearances, rape, threats, extortion, kidnappings, and recruitment of children as combatants. The most common abuses are killings of and threats against civilians, including trade unionists, journalists, human rights defenders, and victims of the AUC seeking restitution of land and justice as part of the Justice and Peace Process. (HRW 2010: 39)

Two of the principal purposes of paramilitary violence, prior to the demobilization of the AUC in 2006, were forced dispossession and the persecution of social movements. Both of these were accomplished through the use of terror-based strategies in the form of threats, harassment, torture, rape, murder and disappearances. The paramilitary used to be identified repeatedly by official sources and human rights agencies as the principal agent in forced displacement and the systematic killings of unionists. If paramilitarism has indeed been dismantled, as Colombian politicians, the state's coercive apparatus, and the media would like us to believe, then one would logically expect that the number of forced displacement and unionist assassinations would have decreased significantly. As I have shown, the situation has been quite the opposite.

As noted earlier, displacement has actually been exacerbated since 2006 as paramilitary operations continue to seize land, which is then transferred into the hands of agribusinesses and mining enterprises or is used for illegal crops cultivation. Land is also appropriated by paramilitaries themselves, as commanders like alias Cuchillo, alias Loco Barera, and alias HH, have been expanding their properties in the Departments of Casanare, Meta, Guaviare, Vichada and Arauca. By 2008, the figures for forced displacement had reached a remarkably high level, according to CODHES. From 2006 to 2007 (during the first year after the demobilization) there was a 38 per cent increase in displacement, and from 2007 to 2008 there was a 25 per cent increase (CODHES 2009). Five years after the demobilization (1 January 2007 to 31 December 2011), already 1,512,405 people had been forcibly displaced (CODHES 2012). In fact, the Internal Displacement Monitoring Centre (established by the Norwegian Refugee Council) declared in its 2007 Global Overview of Trends and Development report, that Colombia, Iraq and Sudan together account for approximately 50 per cent of the world's displaced people. Colombia has also earned the rather shameful reputation of being the world's leader in homelessness (Wilson 2010). Unsurprisingly, land ownership in Colombia continues to intensify – in 2010, 4 per cent of landowners controlled 61 per cent of the best quality land (Verdad Abierta 2010). A study by the University of the Andes has found that the rural Gini coefficient in Colombia is 0.85 (Semana 2010a).

As mentioned earlier, forced displacement is sometimes carried out in urban areas, as the following example from Medellín shows. In the first ten months of 2009, according to the Public Ministry's Human Rights Unit for Medellín (Personeria de Medellín), 2,103 persons were displaced within the city of Medellín, nearly tripling the number of reports received the previous year.

Table 4.1   Cases of Forced Displacement in Medellín in 2009

*Armed Actor Responsible for Displacement**	*Percentage of Cases*
Paramilitary Groups	32
Unidentified Armed Groups	24
Demobilized Paramilitaries	10
Gang members	28
Criminals	4
Army Personnel	1
Guerrilla Groups	1

Source: HRW (2010).
*Based on victims' testimonies.

Nonetheless, land acquisition continues to be the major purpose of displacement and therefore deserves more attention. A considerable part of the paramilitary violence in the post-demobilization era has been manifested in the re-victimization of those who had been violently dispossessed by the AUC in the past. This has occurred through a second wave of displacement (when former victims dare to return to their homes and occupy the properties) or through the selective assassination of and threats against leaders and members of movements for land recovery formed by the victims of paramilitary violence as well as their supporters. Forty-five leaders of various social organizations made up of victims of paramilitary displacement were assassinated between 2006 and 2010 (Semana 2010c) and the headquarters of these organizations were burned down and robbed. Table 4.2 offers some examples. In San Onofre, Department of Sucre, the land-recovery struggles of 52 families who were forcibly displaced from their cooperative farm La Alemania in 2002 by a paramilitary group under Rodrígo Mercado Peluffo turned them into the target of paramilitary intimidations and attacks, resulting in numerous deaths. In 2006 some of the families led by Rogelio Martínez, leader of the Movement of Victims of State Crimes, who had become the representative and leader of the 52 families, returned to La Alemania. Gradually, the rest of the families followed their example. They all worked on the farm during the day but returned to the town of San Onofre at night. In April 2007 one of the farmers was found assassinated. In October of the same year 30 armed men, including some of those who participated in the original displacement, entered La Alemania and threatened the occupants. Subsequently, many refrained from going back to the farm. In July 2008, Martínez began to receive death threats on his phone. Nevertheless, he and some of the other families continued to work on the farm risking their lives. On 24 December 2008 in the town of San Onofre, Martínez was approached by one of the demobilized combatants from the Bloque de los Heroes de los Montes de María who accused him of creating too much trouble for the paramilitary with his talk about stolen land and mass graves. In January 2009, armed men approached Martínez and threatened him again. After Martínez had attended a forum in May 2010 – where other leaders of displaced people, together with NGOs, human rights activists and government representatives discussed the issue of forced displacement – he was shot and killed as he approached the farm building in La Alemania on a moto-taxi. With his death, the number of peasants from La Alemania killed for trying to recover their land totalled 15 (Semana 2010c).

Threats and killings by the Águilas Negras and other present-day paramilitary groups have forced an exodus of families who had been relocated by the government to farms confiscated from paramilitary drug-traffickers under investigation in the Departments of Valle del Cauca, Córdoba, Antioquia, Sucre and Bolívar (Semana 2009a).

In addition to dispossession, the other principal paramilitary activity which continues to take place today is the attack against labour union

Table 4.2   Assassinations of Leaders and Supporters of Victims of Forced
Displacement 2007–2011[5]

Name	Social Position/Occupation	Date of Assassination	Place (Department)
Yolanda Izquierdo	Leader of the Popular Organization for Housing	January 2007	Córdoba
Freddy Abel Espitia	President of the Committee for the Displaced of Cotorra	January 2007	Córdoba
Valdiris Padron	Leader of the Displaced of Pueblo Nuevo de Neclocli	February 2007	Antioquia
Osiris Amaya	Teacher from the Wayuu indigenous group and a defender of the displaced	March 2007	La Guajira
Jose Sosa	Leader of the Displaced in Buenaventura	May 2007	Valle del Cauca
Luis Miguel Gomez	Leader of the Displaced of Montes de María	May 2007	Sucre
Francisco Puerta	Leader of the Peace Community of San José de Apartadó	May 2007	Antioquia
Manuel Lopez	Leader of the Organization for Displaced People	June 2007	Bolívar
Dario Torres	Coordinator of the Humanitarian Zone of Alto Bonito	July 2007	Antioquia
Miguel Orozco	Leader of the Displaced of Tumaco	August 2007	Nariño
Julio Cesar Molina	Member of the Foundation Nuevo Amanecer of the Displaced of Cartago	May 2008	Valle del Cauca
Azael Hernandez	Leader of 27 Forest Protector Families in Tierralta	June 2008	Córdoba
Martha Obando	President of the Association of Displaced Women	June 2008	Valle del Cauca
Alexander Gomez	Leader of the Legion Afecto-Retorno	July 2008	Antioquia
Juan Jimenez	Leader of the Displaced	July 2008	Antioquia
John Correa	Member of the Committee for the Defence of Human Rights	July 2008	Caldas
Walberto Hoyos	Leader of the Community of Curvarado	October 2008	Chocó
Benigno Gil	President of the National Peasant Table	November 2008	Arauca
Carlos Cabrera	Leader of the Displaced of Arauquita	November 2008	Arauca
Jaime Gaviria	Leader of the National Peasant Table	December 2008	Chocó
Alejandro Pino	Leader of the Displaced in Process of Land Recovery	February 2009	Antioquia
Ana Isabel Gomez	President of the Municipal Committee of Families of Victims of the Conflict	April 2009	Córdoba

Name	Social Position/Occupation	Date of Assassination	Place (Department)
Antonio Blandon	Leader of the Association of Internally Displaced Afro-Colombians	June 2009	Bolívar
Jose Betancourt	Member of the Peasant Association of Bajo Cauca	July 2009	Antioquia
Jesus Guacheta	Indigenous leader	May 2009	Cauca
Argentino Diaz	Leader of the Displaced of Curvarado	January 2010	Chocó
Alberto Valdes	Leader of the Association of Victims for the Restitution of Land	May 2010	Antioquia
Rogelio Martínez	Leader of the Movement of Victims of State Crimes	May 2010	Sucre
Alexander Quintero	Coordinator of the Association of Victims of the El Naya Massacre	May 2010	Cauca
Luis Socarras	Wayuu indigenous leader	July 2010	La Guajira
Jair Murillo	Leader of the Association of Internally Displaced Afro-Colombians	July 2010	Valle del Cauca
Beto Ufo Pineda	Leader of the Organization Nueva Florida	August 2010	Cauca
Alvaro Montoya	President of the Board of Community Action of San José de Apartadó	August 2010	Antioquia
Hernando Perez	Leader of the Association for the Restitution of Property and Land of Urabá	September 2010	Antioquia
Edgar Bohorquez	President of the Association of United Displaced People	September 2010	Arauca
Oscar Maussa	Leader of the Cooperative of Farmers of Blanquicet	November 2010	Bolívar
Yonnel Villamil	Member of the Foundation Nuevo Amanecer	January 2011	Tolima
Andres Arenas	Member of the Foundation Nuevo Amanecer	January 2011	Tolima
Eder Verbel Roacha	Leader of the National Movement of Victims of State Crime	March 2011	Sucre
David de Jesus Goez	Leader for the Restitution of Lands in Urabá	March 2011	Antioquia
Ana Córdoba	Member of the Women's Organization Ruta Pacifica de Mujeres and Leader of Displaced Families in Medellín	June 2011	Antioquia*
Antonio Mendonza	Polo Democratico Party Member of the Municipal Council of San Onofre	June 2011	Sucre**

Source: Semana (2011b).
*Source: Sánchez-Garzoli (2011).
**Source: WOLA (2012).

leaders and members. On 5 May 2010 several members of the executive of the Valle del Cauca branch of the Union of Colombian University Employees (Sindicato de Trabajadores y Empleados Universitarios de Colombia, or SINTRAUNICOL), and members of the local branch of the Confederation of Colombian Workers federation (Central Unitaria de Trabajadores, or CUT), received threatening phone calls and email death threats signed by paramilitary groups. On 20 May 2010 all members of the executive committee of the farm workers' union FENSUAGRO received threatening phone calls from the paramilitary. On 18 July 2010 Martha Cecilia Diaz, President of the Santander Association of Public Servants (Asociación Santandereana de Servidores Publicos, or ASTDEMP), received threatening phone calls and email death threats signed by paramilitary groups (ICTUR 2010). Attacks on the labour movement by the paramilitary are not limited to death threats nor are the latter the only form of human rights violations unionists face today. Between 1 January 2007 and 31 December 2011, 218 unionists were killed (USLEAP 2011; El Comercio 2012). The following are a few examples.

On 26 July 2011, Rafael Tobón Zea, a member of the National Union of Workers in Mining, Oil, Agro-combustibles and Energy (Sindicato Nacional de Trabajadores de la Industria Minera, Petroquimica, Agrocombustibles and Energetica, or SINTRAMIENERGETICA), was assassinated in Segovia, Department of Antioquia. On 31 July 2011, Wilmer Serna, a member of the National Union of Agricultural Workers (Sindicato Nacional de Trabajadores de la Industria Agropecuaria, or SINTRAINAGRO), was assassinated in Apartadó, Department of Antioquia. On 31 July 2011, Eduardo Fabián Zúñiga Vasquez, a member SINTRAINAGRO, was assassinated in Apartadó, Department of Antioquia. On 1 August 2011, Over Dorado Cardona, National Secretary of labour issues of the Colombian Federation of Educators (Federación Colombiana de Educadores, or FECODE), narrowly escaped an assassination attempt when armed men attacked him in Medellín, Department of Antioquia. His bodyguard was wounded in the attack. On 30 August 2011, Jailer González, President of the Association of Peasant Workers of Tolima (Asociación de Trabajadores Campesinos de Tolima, or ASTRACATOL), received a series of death threats from paramilitary groups in Rovira, Department of Tolima. On 1 September 2011, José Manuel Espinosa Pachón, President of the National Union of Workers of the National Coffee Federation (Sindicato Nacional de Trabajadores de la Federación Nacional de Cafeteros de Colombia, or SINTRAFEC), received death threats from paramilitary groups in Chinchina, Department of Caldas. On 1 September 2011, Albeiro Valenzuela Soto, a member of ANTHOC, received threatening phone calls from paramilitary groups, while attending a strike by workers protesting against unfair dismissals.

Table 4.3   Number of Unionists Killed 2007–2011

Year	Number of Unionists Killed
2007	39
2008	52
2009	47
2010	51
2011	29*
Total	218

Source: USLEAP (2011).
*Source: El Comercio (2012).

On 9 November 2013, Oscar Lopez Trivino, Nestlé employee and member of National Union of Food Industry Workers (Sindicato Nacional de Trabajadores de la Industria de Alimentos, or SINALTRAINAL) was assassinated following a death threat received by the union on the previous day, signed by the paramilitary group Los Urabeños, and stating 'Guerrillas sons of bitches, keep on pissing Nestle and no more forgiveness you will be chopped to death all the Communists of SINALTRAINAL' (SINALTRAINAL 2013). The attacks against labour unionists have continued into 2014. On 4 January 2014, Ever Luis Marin Rolong, leader of the Brewery Workers Union (Sindicato Nacional de Trabajadores Cerveceros de Bavaria, or SINALTRACEBA) was assassinated.

In addition to labour unions, other social movements have also been targeted by illegal armed groups. On 2 May 2008, the National Ombudsperson warned authorities of a threat that has been circulated through a pamphlet addressed to certain media outlets, NGOs, politicians and social organizations. The pamphlet stated that the following list of people and organizations were a military target: Senator Piedad Córdoba, the Bolivarian Youth Movement, the Anti-imperialist Brigades, the human rights Lawyers Collective José Alvear Restrepo, Corporación Sembrar, the Foundation for Solidarity with Political Prisoners, Corporación Reiniciar, Corporación Yira Castro, Foundation Manuel Cepeda, Asonal Judicial, the Inter-church Commission for Justice and Peace, CUT, Minga, Fundip,

Asomujer, Tao and CODHES. A few specific individuals working with the Movement of Victims of State Crimes, including Senator Ivan Cepeda, were also on the list. At the end of the threat, it was stated clearly: 'We are not emerging gangs. We are the Águilas Negras and are here as an army of social restoration' (Semana 2009b).

There are plenty of examples illustrating how civilians active in social organizing who challenge market-oriented policies and projects promoted by the state and private companies continue to face risks to their lives and safety as they did during the AUC reign. In August 2010 in the municipality of San Francisco, Department of Antioquia, after an academic forum on land and water where various social organizations debated the impacts of the proposed construction of a hydroelectric plant, Águilas Negras sent out flyers to the community stating that their actions 'will focus on the municipalities in Eastern Antioquia where there are still criminals and insurgents'. Moreover, the flyer warned 'addicts, drug-vendors and gossipers' to 'correct their behaviour, otherwise they would be declared a military target' (Semana 2010c).

On 3 January 2014, Mario Arenas Peña, member of the Marcha Politica and defender of displaced people, was killed in Barranquilla, Department of Atlantico. On 4 January 2014, Yovanni Leiton, member of MIA, and his partner Doris Liliana Vallejo were tortured and killed in San Jose del Palmar, Department of Chocó. On 13 January 2014, peasant leaders from the Rural Workers Association of Valle del Cauca (Asociación de Trabajadores Campesinos del Valle del Cauca, or ASTRACAVA), the Departmental Coordination of United Popular of Southwest Colombia (Coordinación Departamental Valle del Cauca del Proceso de Unidad Popular del Sur Occidente Colombiano, or PUPSOC) and MIA received a death threat from the paramilitary group known as Bloque Militar Valle del Cauca – Águilas Negras – Rastrojos (Agencia Prensa Rural 2014).

Healing the Wounds: Prospects for Justice and Reparations for the Victims of Paramilitary Violence

Combined with all the factors discussed so far, impunity has served as a permissive condition for the preservation and consolidation of paramilitary networks. Let us briefly review how the Justice and Peace Law, the extradition of paramilitary commanders to the US, the lack of progress in the prosecution of para-politicians, and the Victims and Land

Restitution Law, have all served to reinforce the image that the Colombian state seeks to bring justice and reparations to the victims of paramilitary violence, when in fact the state is incapable and unwilling to dismantle paramilitarism. Five years after the Justice and Peace Law was approved, only two convictions had been secured. Of the 31,671 demobilized paramilitaries, two thirds were pardoned and did not receive any sentences at all (Verdad Abierta 2012c). From the ten million hectares of land[6] stolen from small-scale farmers by the paramilitary, INCODER has distributed only 894,000 hectares to displaced people (El Tiempo 2012). It should be noted that this does not constitute land restitution since the property distributed is not the same as that which belonged to them, but is rather of poor quality, remotely located, without infrastructure or a terrain that belongs to a national park.

Another example of a legal measure conductive to impunity and the preservation of paramilitary networks was the extradition of 14 paramilitary leaders to the US in May 2008, where they are in the process of being prosecuted for drug-trafficking. What is presented as a punishment is in reality a move to protect both parts of the terror apparatus: the paramilitary chiefs and the Colombian state agents involved in paramilitary operations. Instead of being investigated for crimes against humanity and war crimes, including some 200 massacres, the disappearance of at least 49,000 people, and numerous cases of torture, beating, mutilation, rape, recruitment and abuse of children, and so on, they are prosecuted for the crime of drug-trafficking. As a result of these extraditions, neither the Justice and Peace Law nor ordinary Colombian justice can be applied to those responsible for mass human rights violations; they are also out of the reach of the International Court of Justice because the US does not recognize its jurisdiction (Hristov 2009a).

The presence of paramilitary power at all political levels has been so pervasive, as illustrated earlier, that despite the investigations of paramilitary politicians underway, the Colombian state is unlikely to cleanse itself of paramilitarism anytime soon. There are four institutions in charge of investigating the links between paramilitaries and politicians: the Supreme Court of Justice, the Fiscalia, the Attorney General's Office (Procuradoria General de la Nación) and the Commission for Accusations against Members of the Chamber of Representatives. But there is very little collaboration between these four institutions. Through the confessions of some of the demobilized paramilitary commanders, 11,179 politicians, government officials and entrepreneurs have been implicated with paramilitary groups. Yet, there have been very few convictions. For instance, in the Department of Antioquia, one of the departments with the most para-politics, by 2012, 22 former Congress members had been on trial but only seven have been sentenced. Of eight politicians known to have committed crimes against humanity, particularly for having participated in massacres, only two have been sentenced. Due to the lack of resources and personnel, investigations proceed very slowly and many cases result in a statute of limitations (*vencimiento de terminos*) as a result of which the case is closed without any further judicial process taking place. According to judicial investigators, the businesses that resulted from para-political alliances remain intact. The capture of the principal leaders does not affect the structures. Even if a Congress member (as co-owner of a business) has been sentenced, the enterprise is not investigated or closed down since there is no specialized unit in the Fiscalia to investigate the paramilitary economy (Verdad Abierta 2012b).

Lastly, let us look at the Victims and Land Restitution Law[7] (Ley de Víctimas y Restitución de Tierras, Law 1448), which President Juan Manuel Santos signed on 10 June 2011. One of the most important and most highly debated issues covered by the law is that of land restitution. The law is presented as a tool that will help in the restitution of millions of hectares of lands stolen or illegally appropriated through human rights violations, and the return of these to the rightful owners. There is considerable scepticism among some on the Left in Colombia about this promise. It is my belief that the Victims and Land Restitution Law is just another strategy to divert attention away from the fact that paramilitary groups continue to engage in dispossession and freeing up of land for the expansion of agribusinesses and mining operations by local and foreign capital. I will provide here a short summary of the key shortcomings of this law.

To begin with, the law considers only two million hectares of land to have been illegally appropriated and there is a big discrepancy between this figure and the estimate by other social organizations and movements, such as MOVICE, who put the figure at ten million. Secondly, the victims eligible for land restitution under this law include only those who lost their land after 1991. Clearly, this excludes hundreds of thousands of other victims who were dispossessed in the 1980s. Thirdly, victims are expected to provide precise information related to their land registration but a large number of them no longer have such documents after having to flee their homes in terror and living in precarious conditions for many years.

Fourthly, the law requires the victims to travel to the region from which they were displaced in order to present their case to the judge. Many are too frightened to go back, given the continuation of paramilitary attacks against land-recovery initiatives. Thus, it is very likely that many people will be impeded from reclaiming their land.

As demonstrated earlier, often the land has passed from the hands of the original expropriators (the paramilitary) to *testaferros* or companies or individuals who may well claim to have bought it in good faith, making the process of establishing the rightful owner very difficult. In addition, as a 2012 report from Amnesty International explains:

> If someone is in possession of land and has developed it, for example for agro-industrial production, they will only be required to pay a rent to the rightful owner, unless it can be proved that they stole the land in the first place. As the vast majority of demobilized paramilitaries are not under investigation for human rights violations, it is extremely unlikely that many people who were acting in collusion with them or on their behalf will be exposed, so proving that occupiers acted illegally is both difficult and rare. (AI 2012: 16)

Only in rare cases agro-industrialists who acquired land that was made available through paramilitary operations have been sentenced, and in those cases the sentences were quite short. For example, two palm oil entrepreneurs, Luis Alberto Flórez Pérez and Iván Patiño Patiño, who were allied with AUC groups responsible for displacing thousands of Afro-Colombians in the Department of Chocó, received prison sentences of between four and five years (Territorio Chocoano 2011). Another interesting point that the Amnesty International report makes is that:

> Illegal armed groups have frequently brought in civilians from the area and from further afield to work on agro-industrial projects on stolen lands. Many of these workers are forcibly displaced and the Law could result in the recognition of claims by these workers over the lands they currently work, possibly at the expense of others who were forcibly displaced from these areas ... The Law does not appear to provide full support to peasant farmers wishing to return to their lands and engage in subsistence agriculture. Rather, the emphasis appears to be on encouraging peasant farmers to participate in agro-industrial, infrastructure, tourist or mining projects. In effect, Article 99 could see

the continuation of some agro-industrial projects which were either the reason for the forced displacement in the first place, or were initiated in the wake of land theft resulting from human rights abuses. (AI 2012: 15)

The law stipulates that those who qualify for compensation are the victims of the armed conflict. This leaves one wondering about those who have been re-victimized through new waves of displacement in the post-demobilization era. Such people would not be recognized as victims of the armed conflict but only as victims of BACRIM (criminal gangs), and therefore would not be eligible to make claims under this law. To conclude, the Victims and Land Restitution Law was never truly intended to bring about reparations to the victims and reverse the counter-agrarian reform under way in Colombia. The single most important reason is that the law is based on two false premises: that the armed conflict is a thing of the past and that the current armed groups are not paramilitary. Thus, the state renders invisible the fact that neither the armed conflict nor paramilitarism is over. The problem of forced displacement and land expropriation cannot be resolved as long as the continued existence of the problem is denied. To believe that the harms of the past can be repaired by the same unaltered system that caused those harms is completely absurd.

5

Paramilitarism, Neoliberalism and Globalization: Towards a New Analytical Framework

The way the phenomenon of paramilitarism is theorized can determine how the problem of existing non-state armed groups is perceived and consequently dealt with. Throughout this book my goal has been to properly identify and analyze a persistent and evolving entity implicated in processes of capital accumulation, the creation of inequality, the use of irregular armed force, and the perpetration of human rights violations in an officially democratic country. If the political and economic linkages that paramilitary groups exhibit simultaneously in relation to society and the state are obscured then the term 'paramilitary' becomes quite meaningless and useless. The common usage of the term so far has served to erect a conceptual barrier in two ways. The paramilitary has been seen as either attached to the state only (and not to the capitalist classes) or as associated with a certain sector of society (but not the state). When perceived as connected to the state, the paramilitary is conceptualized as a subservient extension of the military and the connection is not seen to extend to any other state institutions outside the coercive apparatus. When viewed in relation to a specific social sector, the paramilitary is overwhelmingly characterized as a violent agent whose only purpose is to enable a particular illicit economic activity (i.e., drug-trafficking) rather than a wide range of economic schemes that form part of the neoliberalization of the economy as a whole, such as agribusinesses, resource extraction industries and labour deregulation. To break out of these counter-productive conceptual confines and to be able to adequately capture the emergence and development of paramilitarism in Colombia, I propose an anti-ideological analytical framework based on three guiding principles.

First Principle: Paramilitarism as a Multidimensional Entity

Paramilitarism is a multi-dimensional phenomenon that is simultaneously economic, political and military in nature. The expression that each of these dimensions takes is historically contingent. Throughout Colombia's history, paramilitary operations have included persecution and combat of guerrilla forces, attacks against civilians perceived as left-wing and/or guerrilla sympathizers, violent dispossession through forced displacement, and the running of criminal enterprises. The paramilitary has had a political presence at all levels of political institutions, as illustrated in Chapters 3 and 4, and has consistently supported right-wing pro-neoliberal politicians. Thus, paramilitary activities spill over the political realm[1] of right-wing ideology associated with maintaining the status quo, the military realm constituted by the use of armed force to murder civilians and combat guerrilla forces, and the economic realm through the appropriation of land, as well as the operation of legal and illegal businesses (see Figure 5.1).

The presence of the paramilitary in each of the three types of realms is made possible through the use of violence for an economic end. For instance, the election of paramilitary friendly politicians is accomplished through campaigns of intimidation of voters and assassinations of opposition candidates. The significance of having politicians who represent the interests of the paramilitary is that economic policies, legal procedures around land registry, and military security projects all facilitate capital accumulation by those who employ paramilitary forces. Similarly, if we look at the military realm, the combat of the guerrilla, which implies the use of violence, is done with the aim of protecting the security and interests of capital. Those who accumulate capital, whether through legal or illegal means, see peasant movements, labour unions and left-wing educators, for instance, as obstacles to the advancement of their interests as much as they see the guerrilla as an enemy, and consequently resort to violence whenever necessary to neutralize the threat. Hence, what appears to be a dichotomy between a political movement (as Romero [2003] saw the paramilitary in the past) versus a criminal organization that uses violence to gain wealth (as Duncan [2006] defined it) is false, since the question of whether the paramilitary is after political gain or material assets has to do with its strategy at a particular point in history, but not its underpinning motivation (that is, ultimate goal – enrichment), which has

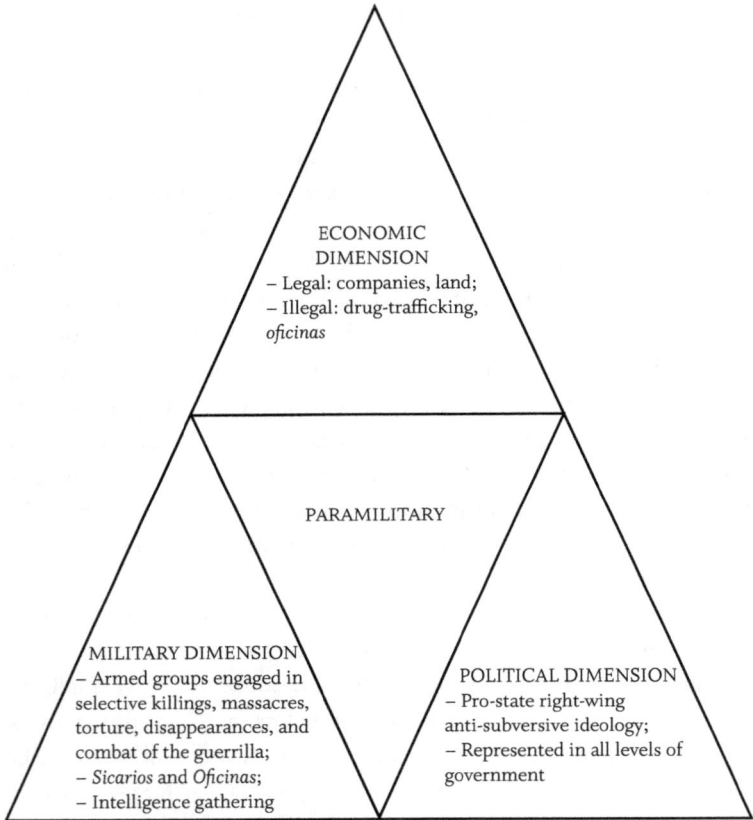

Figure 5.1   Economic, Political and Military Dimensions of Paramilitarism

remained consistent throughout history. Exercising power and influence over political decision-making is a means to an end. As Miliband explains:

> More than ever before men now live in the shadow of the state. What they want to achieve, individually or in groups, now mainly depends on the state's sanction and support. But since that sanction and support are not bestowed indiscriminately, they must, ever more directly, seek to influence and shape the state's power and purpose or try to appropriate it altogether. It is for the state's attention, or for its control, that men compete. (1973: 3)

This principle, which enables us to recognize the multi-dimensional nature of paramilitarism, can help us challenge the Colombian state's ideological construct of BACRIM, which rests on the characterization of the paramilitary (by some academics) as a criminal organization. The ideological support for this claim is that paramilitary units used to have the political objective of defeating the insurgency, while illegal armed groups today are interested only in illicit enrichment through drug-trafficking, extortion and other illegal activities. Drawing on Bannerji (2001), an anti-ideological analysis is not about replacing 'false' content with that which is 'true', but about looking at the social relations shaping knowledge production. It is true that illegal armed groups today carry out extortion, drug-trafficking and other illegal operations. Nevertheless, at the heart of the ideological production here is a conjuring act whereby a multi-faceted phenomenon is viewed in terms of one expression only at a given moment in time. Subsequently, these different expressions (from different time periods) are contrasted with one another and it is concluded that what exists today is fundamentally different from what existed in the past. A closer look at the recent history of paramilitary development reveals how this is done.

Intersection of Paramilitarism and Organized Crime

Prior to the demobilization, the paramilitary's sources of funding encompassed criminal activities. They worked with and in many cases owned (the way one can own a franchise) criminal organizations which engaged in the smuggling of gas and weapons, extortion, robbery of security vehicles, car theft, condominium and house robberies, and the operation of casinos, brothels and live sex show establishments. La Banda de los Calbos in Cali and La Banda de la Terraza in Medellín are two examples of criminal organizations that worked for the paramilitary. For instance, La Banda de la Terraza, an organization which consisted of 300 men, engaged in stealing cars and re-selling them, robberies, kidnappings and the murder of unionists and other individuals suspected of being Communists (Interview 2005). According to a former military officer, who had dealings with La Banda de la Terraza, the latter used to provide the paramilitary with stolen cars which they used to drive the corpses of the victims to the mass graves (Interview 2007). The *oficinas de cobro*, which can be described as illegal collection agencies that operate through death threats, kidnappings and murder, were also among the illegal enterprises they ran. The Bloque Cacique Nutibara is one example of a paramilitary group that comprised criminal gangs and *oficinas de cobro*.

Those who work for the *oficinas* are *sicarios*. It is not uncommon to find active members of the police and military participating in such ventures. It is in examples like this that we see the intersection of paramilitarism with organized crime. Yet this criminal aspect of the pre-demobilization of the paramilitary has remained mostly unremarked. One of the reasons for this was that the AUC used to present itself to the national and international community as 'a political-military movement which uses the same irregular methods as the guerrillas. Its members are not terrorists, nor common criminals' (Pizarro 2004: 120). The AUC portrayed itself as an anti-insurgent armed movement which was not criminal in nature and did not seek private interest but rather had collective ideals. The purpose of the social discourse of the AUC was to gain a political recognition as a legitimate force. Some academics, such as Romero (2000), considered the paramilitary to be a political actor because of its origin and counter-insurgency projection. Moreover, in order for paramilitary commanders to receive all the generous benefits under the Justice and Peace Law (after the demobilization process), such as a mere five to eight years in jail for perpetrating massacres, the paramilitary had to be perceived as an armed organization with a political agenda. Emphasizing their criminal characteristics would only have damaged this image. In fact, there were drug-traffickers who bought the AUC 'franchise' in order to gain legitimacy as a political force and disguise their criminal character, and subsequently allow them to take advantage of the benefits offered by the Justice and Peace Law.

The situation is different today, because the interests at stake have changed. Distancing the state from the paramilitary has become a priority for the former. Since too much has already been revealed about the state's complicity in paramilitary human rights abuses, it is undoubtedly more convenient to produce a body of ideas that create the impression that all such connections, no matter how numerous and profound, belong to the past. Such a discourse can only rest on the claim that paramilitary groups no longer exist today. To this end, the Colombian state and media pose the categories of 'drug-trafficker versus paramilitary' or 'criminal versus anti-subversive', as being necessarily mutually exclusive (as a number of authors discussed in Chapter 2 had presented them), thus creating a false dichotomy. In other words, the claim is that the armed organizations of the past were paramilitary because they were anti-subversive while the armed forces of today are criminal because they engage in drug-trafficking. One kind of expression of paramilitarism from the past (military realm

– combating the guerrilla) is compared to a different kind of expression (economic realm – drug-trafficking) in the present. One of the advantages of such an ideological manipulation is that the Colombian state no longer has the same degree of responsibility to provide support to the victims of forced displacement or other forms of human rights violations carried out by present armed groups since these are (according to the state's definition of BACRIM) no longer related to the country's armed conflict. While criminal gangs do exist in Colombia, to reduce present-day illegal armed groups to 'criminal gangs' is a gross distortion of reality. Criminal gangs do not have the kind of solid relationship with judicial, political and military state institutions that is necessary to secure impunity for their crimes, provide them with ammunition, facilitate their operations, and convert illegally accumulated wealth into legal capital. Even 'organized crime' is not an adequate term to describe these groups.[2] Most criminals cannot expect systematic collaboration from state institutions like the military, police, DAS and INCODER that provide paramilitaries with the security and legitimacy to ensure their operations are successful. Criminals also do not share a common enemy with the state (that is, the guerrilla) and do not typically attempt to exterminate the urban and rural labour movement. Clearly, it is time to think about how to conceptualize the paramilitary as a violent actor with a relationship to the state, the capitalist classes and processes of land acquisition and capital accumulation in general.

The Murky Waters of Drug-Trafficking and Paramilitarism: Establishing Conceptual Clarity

While drug-trafficking is just one of the illegal activities that fund paramilitary groups, it deserves special attention not only because it has been the most lucrative one but also because it is has given rise to competing claims about who the paramilitary really is. While the current ideological production by the state positions drug-trafficking (criminal) and paramilitary (political) within a false dichotomy in order to emphasize a contrast between past and present, there has been another kind of distortion when it comes to conceptualizing the paramilitary. Many have been tempted to conflate the paramilitary and criminal organizations into the same category of illegal armed groups or non-state armed actors, obscuring the crucial difference between the two. Before I proceed, it is necessary to pause and briefly look at the political economy of drug-trafficking. It involves a wide range of actors: those who cultivate the coca plant (poor peasants) on their own land or that owned by the paramilitary; the *raspachines* (those who pick the plant); the *cosineros* (those who work in the processing facility where chemicals are added to the dry plant to turn it into coca base); the *empacadores* (those who pack it); security personnel who guard the laboratories; those who transport it to a sea port or to an airport; the *cobradores* (those who settle unpaid accounts), the *sicarios* (who usually carry out assassinations, kidnapping, or extortion), *hombres de confianza* (the men of trust, also derogatorily called *lava-perros* or *lugartenientes*, which is the term the police use to refer to them); *el patron* (the drug-lord) who is the owner of a given route (this refers not only to the physical space through which the drugs are transported but also to the established arrangements and people along the way who make the trafficking possible); and finally the distributors and sellers abroad.

Many have pointed to the fact that since all these actors operate within a sphere of illegality characterized by the absence of legal regulatory mechanisms, they cannot rely on law enforcement when conflict arises the way legal enterprises can. Thus, they resort to para-legal or violent mechanisms to enforce 'the rules' and punish those who break them. So, for instance, the provision of security around *cosinas*, the vehicles transporting the illegal drugs and money, the collection of unpaid debt, the elimination of competitors/rivals, and confrontations or negotiations with the police, all require the service of armed individuals. Inherently then, drug-trafficking is not a peaceful economic activity, but rather one dependent on violence or the threat of such in order to be successful. The armed individuals or groups who perform these violence-based services to enable or facilitate drug-related operations (commonly referred to as cartel violence) are *not* paramilitary groups. As Mazzei (forthcoming) explains, drug cartels are criminal economic ventures. Their ultimate interest is their own criminal enterprise. Arias defines criminals as 'actors who regularly engage in illegal activities as a primary vocation but who have no formal ties to the state', while acknowledging that there are often actors within the state who engage in illegal activities (2006: 12, cited in Mazzei, forthcoming). The targets of the violence enacted by the armed actors working on behalf of the drug-trafficking organizations in Colombia include anyone perceived as a traitor, anyone who endangers the operations of the organization, competitors, those who do not fulfil their obligations (e.g., those who fail to pay money owed), and so on. This does not constitute paramilitary violence.

As mentioned earlier, drug-traffickers were among the founders of many paramilitary organizations that emerged in the 1980s. Why did they become implicated in paramilitarism and in what ways? In the 1980s many drug-lords began to invest their capital in land, usurping millions of hectares of the best agricultural land in the country across regions such as Magdalena Medio, Ariari, Urabá, as well as the Departments of Córdoba, Risaralda, Caldas, Quindio and Valle del Cauca. Thus, drug-traffickers came to form a part of the rural oligarchy – the landowners. Of course, drug capital was also invested in a range of other kinds of property and commercial activities. The FARC represented a threat to the interests of drug-traffickers as it did to any other members of the capitalist class. This threat took the form of kidnappings, forced taxation, and of course the potential land reform and wealth redistribution that could take place if the FARC were to take over state power. Paramilitary bodies were created to protect the collective interests of these capitalist classes by providing security, combating the guerrilla and exterminating potential guerrilla sympathizers or 'clearing the ground of subversives'. For instance, in 1981, the kidnapping of Martha Nieves Ochoa, whose brothers were members of the Medellín drug cartel, was the impetus for the emergence of the armed group called Death to Kidnappers (Muerte a Secuestradores, or MAS) based in Medellín. It was formed by 223 drug-traffickers as well as some active members of the army and police. As Contreras (2002) explains, the movement adopted the ideology of anti-communism, thus allowing its constituents to establish more solid connections with sectors of the armed forces, given that they were faced with the same 'internal enemy'.

In the 1990s, drug-traffickers (who were also already landowners), along with other members of the capitalist classes, began to employ paramilitary groups not only to attack sectors of the civilian population perceived as a fertile ground for FARC's indoctrination, but also to displace peasants from their land, making possible the transfer of land into their hands or those of agribusinesses or other companies. The important thing to remember here is that drug-traffickers are capitalists, whether or not they engage in any other wealth-generating activity in addition to drug-trafficking. In 2007, the paramilitary drug-traffickers' collective annual income was estimated to be around four billion dollars (4 per cent of the national GDP) (Richani 2007). The interaction of drug-trafficking with paramilitarism thus arose out of the need for drug-traffickers to protect and advance their interests against challenges to their capitalist power. Are there ways in which paramilitary violence has been directly functional

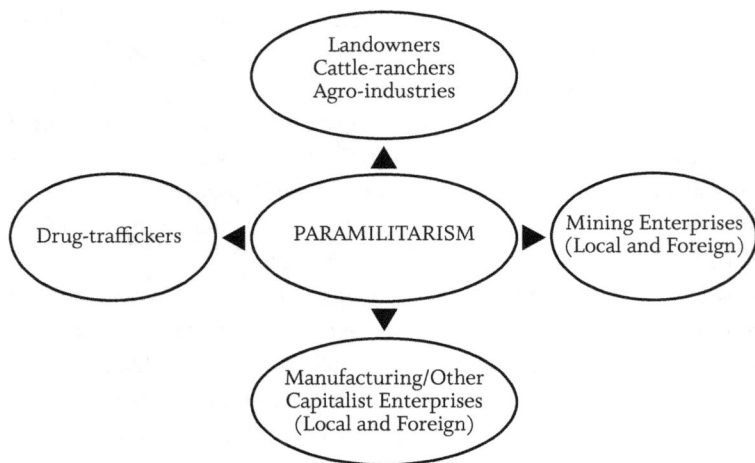

Figure 5.2   Paramilitarism Serving the Interests of Various Sectors of the Capitalist Classes

to drug-trafficking specifically? The answer is yes, in the following way: by organizing, controlling and managing the growth of illicit crops by *campesinos*. For instance, testimonies from Puerto Lleras, Department of Meta, indicate that when the paramilitaries arrived in the rural areas of this municipality they declared to the citizens that those who wished to remain living there must begin to plant coca. The peasants were told that if they produced what was required, the paramilitary would protect them and the state forces would tolerate their work (CINEP 2005b).

Drug-trafficking, like the other illegal activities mentioned earlier, is a source of funding for paramilitary groups who in turn defend the capitalist interests of drug-traffickers. Therefore, drug-trafficking in itself is not the end objective of paramilitary groups. It is rather an activity that generates the wealth needed to sustain them so that they can engage in operations that facilitate capital accumulation mainly through repression of labour or dispossession. To simplify then, when an armed group protects the interests of a drug-trafficker as a capitalist by attacking the guerrilla, social movements challenging the capitalist system, and by displacing people from land in which the drug-trafficker wants to invest, then this armed group should be called a paramilitary group (regardless of whether or not it labels itself as such). Unlike armed groups dedicated strictly to the cartels' operations, paramilitary groups exist not to sustain a particular illicit activity, but rather the entire existing politico-economic system. Of course, there are instances where the same individuals or armed groups that engage in politically motivated (that is, paramilitary) violence also carry out criminal (that is, cartel) violence. This is why it is essential to avoid two kinds of fallacy: one is to equate paramilitarism with drug-trafficking; the other is to conceptualize the two as mutually exclusive categories of violence.

A Word of Caution About the So-called 'Colombianization of Mexico'

The argument that Mexico is becoming a second Colombia was frequently heard about ten years ago from politicians in the North and some academics such as Carpenter (2005) who mainly drew parallels in terms of the gruesome violence, corruption and weak state institutions as well as some of the policies undertaken by both countries to combat these issues. This claim has, however, provoked a reaction among several scholars, such as Scherlen (2009), who felt that it was necessary to differentiate the mostly drug-cartel violence in Mexico from the varieties of violence in Colombia including drug-cartel, but also paramilitary and guerrilla, which are politically motivated. Yet, in January 2014 headlines announcing the formation of paramilitary groups in Mexico, such as *Semana's* article entitled 'These are the Paras', once again invited comparisons of Colombian and Mexican violence. Indeed, during the second half of 2013 in the state of Michoacan, Mexico, groups of armed civilians calling themselves 'self-defence' (*autodefensas*) had organized to defend themselves against the abuses of the drug-cartel Caballeros Templarios whose members have engaged in killings, kidnappings, extortion, rape of young girls and forced taxation imposed on vendors, agricultural producers and cattle-ranchers. According to *Semana* (2014a), these groups were formed in mid 2013 and by August had five municipalities under their control (that is, free of cartel abuses). The state had tolerated their existence until January 2014 when confrontations between Caballeros Templarios and *autodefensas* intensified and state forces had to intervene militarily to impose order. The state then began a dialogue with the *autodefensas* with the intention of jointly coordinating future actions in order to defeat the drug-cartel.

Semana (2014a) raises the question of whether this might not be the embryo of paramilitarism, given that such 'community police' groups are spreading across Mexico, currently present in at least 13 states, totalling over 9,000 men, and supported by 60 per cent of Mexicans. On the other hand, Fernandez (2014) critiques this comparison between Mexican *autodefensas* and Colombian paramilitaries, pointing to the fact that the former, in the case of Michoacan, have returned 654 acres of land seized by Caballeros Templarios to villagers and have forced the mining company Minera del Norte (which used to pay protection money to Caballeros Templarios) to cease its regional operations. She stresses the fact that these *autodefensas* (not only in Michoacan but also other states) protect the interests of local communities against the reign of drug-cartels. That reign is fortified by the rampant corruption including high-level collaboration between cartels and members of the army, police force and government. In some cases the *autodefensas* have also fought against the operations of mining companies (Fernandez 2014).

All these factors make this quite an interesting case and I would warn against hasty conclusions. I agree with Fernandez that at this point, the *autodefensas* of Michoacan cannot be categorized as paramilitary. Instead the term 'vigilante' (see the discussion in Chapter 2) would be a much more appropriate label for them. However, one must also be attentive to how their relationship with the state might develop in the future. On one hand, the *autodefensas* are underlain by the idea of 'community policing' or neighbourhood watch, which has in fact been a long tradition among some indigenous groups. Nonetheless, if we look back at the case of Peru for instance, the *rondas campesinas* were eventually armed professionally and integrated into the state's counter-insurgency strategy. So it will be interesting to see how much (if any) cooperation there might be between state forces and the *autodefensas* in the future, and whether such cooperation will be strictly limited to the elimination of the drug-cartels or will acquire new functions. If the latter occurs, it would be important to note whether their operations have any class-bias (that is, can be linked to protecting the class interests of certain societal sectors). Only then might we be in a position to clearly discern the character of such parainstitutional formations.

What can be safely concluded so far from the Colombia-Mexico comparison with regard to state and parainstitutional violence is the following: General commonalities between the two countries do exist with regard to the use of violence to protect the interests of the capitalist classes. One manifestation of this is the *guardias blancas* working to protect individual landowners as well as the emergence of paramilitary groups in Chiapas and to some extent in Guerrero and Oaxaca. Another broad commonality is the use of the military for internal security and the ways in which the militarization that is part of the War on Drugs serves the interests of mining companies by repressing social activism. This is actually a trend currently underway in Central America. The notorious involvement of state forces in human rights violations, such as torture, forced disappearances and extrajudicial executions, as well as drug-trafficking, combined with a high degree of impunity, constitutes yet another similarity between the two countries. There are however some important differences, in addition to the ones discussed in relation to the developments in Michoacan. Mexico's paramilitary violence is limited to a few states in the South, mostly Chiapas and to some extent Oaxaca and Guerrero. As mentioned earlier, the prevalence of cartel violence in Northern Mexico should not be compared to the cross-country paramilitary violence in Colombia. Armed groups should not be objectified and defined primarily or solely in terms of their access to means of violence. Armed groups are socially embedded agents who through their use of violence seek to influence existing social hierarchies in some way or another. Therefore, it is imperative to distinguish the different nature of parainstitutional formations by identifying: 1) their end goals, 2) those who benefit from their actions, and 3) those who are victimized by them. In Colombia there has been, for the most part, a symbiotic relationship between cartels and paramilitary organizations.[3] In Mexico, research on paramilitary groups in the southern states has not so far demonstrated any such connection.

Colombia: Representative of the Region as a Whole or an Anomaly?

Another commonly asked question that stems from such comparisons is this: If there is poverty and class inequality across much of Latin America, why has Colombia experienced such a persistent armed conflict that has resulted in extensive violence, both quantitatively and qualitatively speaking, as well as rampant human rights abuses, making this country stand apart from all its Latin American neighbours? The seeds of violence at the service of capitalist development were sown in Colombia, as they were in the rest of Latin America, 500 years ago. However, the permissive conditions that allowed these seeds to bear fruit have been present in Colombia to a much larger extent than in the rest of the region. One of these conditions has been the very close relationship between the state and the elite from the beginning of the independent nation-state.

The Colombian elite has not been characterized by internal class-based conflicts or fragmentations. Even though there was a war between Liberals and Conservatives, their divisions did not correspond to any sectional differentiation within the capitalist class, such as landowners versus industrialists. Thus, the elite always stood firm when it came to recognizing challenges to its interests as a class and any conflicts within it did not prevent it from standing united when it came to dealing with threats coming from organized labour and peasant militancy. This allowed the fusion between the elite and state institutions to consolidate. Another factor that distinguishes Colombia from other Latin American nations is the presence of an armed insurgency, known to be the largest and longest-standing in the hemisphere. The fact that Colombia is an economically strategic territory for capital due to its natural wealth including minerals, gold, biodiversity, petroleum, fertile land and precious woods, together with the threat to the security of capital coming from the existence of an armed Leftist movement, is a unique combination that is ripe ground for rapacious violence in the name of capital. This, however, would not have been possible without 'generous' US involvement, making Colombia rank for a long time as the largest recipient of US military aid in Latin America. Finally, the penetration of the economy by drug-trafficking capital and the subsequent strengthening of the economic power of the local capitalist class has been the other significant permissive condition. It is my contention that none of these on its own would have produced such a destructive outcome, but the constellation of all these factors have turned Colombia into one of the most dangerous countries in the world to be a Leftist, if not the most dangerous. As mentioned earlier, paramilitarism exists in different parts of Latin America and its level of development varies from place to place – in some countries it is still in its embryonic stage while in others, such as Mexico, Haiti and Guatemala, it has advanced to a more developed phase. However, only in Colombia has this violent strategy for the purposes of dispossession and repression reached such a high level of coordination and organization across time and space. Nevertheless, the decentralization of policing and military operations through public-private partnerships, currently underway in other parts of the region, is a significant permissive condition that may serve as an impetus for the further development of organized violence of a paramilitary nature. Whether this will come to resemble the Colombian case remains to be seen.

Second Principle: Paramilitarism as a Structural Social Formation

Paramilitarism is rooted in the economic and political structure of Colombian society. As capitalism evolves, so does paramilitarism. A structural phenomenon implies long-term or persistent patterns of relationships. Paramilitarism has been a proactive instrument (to borrow Robinson's [2004] phrase) for the advancement of capitalism through violent dispossession and repression. This means that paramilitarism as such is not limited to a specific armed group. Some paramilitary groups may disband or disarm while new groups may emerge. Regardless, paramilitarism as a phenomenon will persist for as long as there is a fertile ground for it and, for now, the latter is to be found in abundance. One of the central features of the current social landscape, the importance of which cannot be overestimated, is the exacerbating wealth inequality, the causes of which were discussed in the previous chapters. The efforts of the poor to improve their situation take two forms which are favourable to the formation and sustainability of groups of a paramilitary nature. On one hand, the joint struggles of many low-income citizens have given rise to diverse popular movements, which represent a challenge to the status quo and are therefore a target for repression by the paramilitary. On the other hand, given the high levels of unemployment, subsistence opportunities limited to the informal sector, and low incomes on which it is difficult if not impossible to survive, people from the poor sectors of society are willing to do anything in order to move upwards on the economic ladder and attain something resembling the lifestyle enjoyed by the rich. Many of them see no other option but to resort to common crime, while others join well-developed organizations dedicated to theft, extortion, kidnapping and drug-trafficking. Yet others join directly paramilitary organizations. As we can see, capitalist development in Colombia produces conditions that require violent measures and at the same time provides the human resources to enact them (Hristov 2009b).

Neoliberalism and Paramilitarism: A Vicious Cycle

The relationship between paramilitarism and neoliberalism illustrates the cooperation between the paramilitary's economic and military structures. Paramilitarism has served an important function in the neoliberal restructuring of Colombia, as illustrated earlier. It has helped to implement neoliberal policies or has facilitated activities promoted by neoliberal politicians primarily in the areas of agribusiness and resource extraction by 1) displacing rural residents and providing security to companies that take over their land, and 2) attacking labour unions and other movements that rise up against neoliberal policies such as privatization and labour deregulation. The increasing poverty and insecurity, which are the outcomes of neoliberalism, mean that more human resources are available to be channelled or recruited into paramilitary networks that can in turn displace more people and facilitate more capital-friendly development and so on. This is why I argue that the paramilitary is the embodiment of the use of violence for purposes of capital accumulation through its two basic functions: dispossession and repression. As Richani (2007) argues, the neoliberal economic model is consistent with the class interests of most bourgeois factions, including the narco-bourgeoisie, with its ideological orientation that calls for free markets. Thus, according to him, the paramilitary, the narco-bourgeoisie and their political allies form a 'reactionary configuration' which applies labour-repressive practices in agribusinesses and the other economic activities they operate.

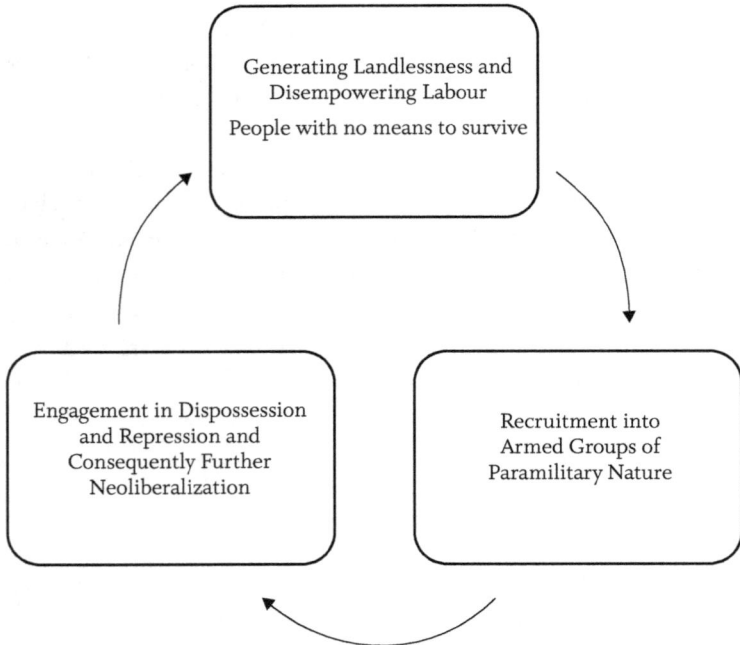

Generating Landlessness and Disempowering Labour
People with no means to survive

Engagement in Dispossession and Repression and Consequently Further Neoliberalization

Recruitment into Armed Groups of Paramilitary Nature

Figure 5.3   The Relationship Between Paramilitarism and Neoliberalism

What About the Guerrilla? Peace-talks Between the Colombian Government and the FARC (2012–2014)

I have been accused at times of being silent about the guerrilla's role in the armed conflict and only focusing on 'one side of the story'. Obviously, those who are under this impression have missed an important detail: my work has set out to unmask and explain capitalist violence, not revolutionary violence. Nonetheless, it might be useful to comment here on the negotiations in Havana between the Colombian government and the FARC (which have been underway since 2012) in light of the para-militarization of Colombia. Comprehending paramilitarism as a structural phenomenon can illuminate greatly the prospect for peace in Colombia, even though the end of the armed conflict is believed by many to depend entirely on the final outcome of these peace talks. In fact, during the CELAC summit in Havana in January 2014, President Santos stated: 'If you ask me whether I am more optimistic now than I was a year ago ... yes, I am more optimistic because I see that there is a will from both parties. On my part, the will is totally there ... I hope that in the next [CELAC] summit, in Costa Rica, we can say that the armed conflict in Colombia is over' (Cabrios 2014). This statement is an expression of two underlying beliefs which are fundamentally wrong: 1) that an armed conflict exists because there is a guerrilla movement, and 2) that the guerrilla is the only actor in the armed conflict outside the state. According to this view, paramilitary organizations no longer exist; any non-guerrilla armed groups are simply criminal gangs which the state is gradually and successfully eradicating. In reality, the guerrilla was not the cause of the Colombian conflict but rather one of its symptoms, and simultaneously became a contributing factor in the sense that its very existence has provided the ideological substance for the pretext and justification behind state-sanctioned violence and militarization. Thus, unfortunately the presence of the guerrilla has been used by the powerful to legitimate the onslaught on social forces that challenge the power of the dominant classes.

If we look at the demands that FARC has presented at the peace talks in Havana, it becomes clear that regardless of how 'willing' the state might appear to be to meet them, such promises cannot possibly materialize in the long run for as long as neoliberalism is the politico-economic model currently in place, fortified by paramilitarism. Let's first briefly review some of the FARC's key demands. The movement believes that peace with social justice can be accomplished through an integral, democratic, participatory and environmentally friendly programme for rural development. Its main points include:

1) securing food sovereignty for the country as a whole;
2) supporting and empowering small-scale and medium-scale farming;
3) eliminating hunger;
4) eliminating landlessness;
5) limiting the foreign ownership of land; and
6) revising the free trade agreements.

Some of the more specific points are:

1) making constitutional the right to food sovereignty and nutrition as a fundamental human right;
2) creating a programme for hunger eradication – a percentage of the GDP has to be spent on food subsidies for satisfying the basic needs of the poorest people until reforms are put in place and free land titles are granted to landless peasants, agricultural workers, poor urban residents and women;
3) creating Peasant Territories for Food Production (Zonas de Produccion Campesina de Alimentos) which will be protected from extraction industries. This includes the creation of 59 ZRCs;
4) creating a Land Fund/Reserve (Fondo de Tierra) which can distribute land to landless families by confiscating idle *latifundios*, illegally appropriated land and land owned by drug-traffickers;
5) developing programmes that facilitate small and medium-scale farmers' access to local markets by improving transportation infrastructure while reducing the role of intermediaries;
6) ensuring a balance between the land used for peasant agriculture and that taken up by agroindustries and cattle-ranching;
7) respecting the autonomy of indigenous groups and Afro-Colombian communities;
8) creating various peasant economic organizations such as a National Council for Food and Nutrition comprising members of peasant organizations; and
9) promoting scientific development and protection of native seeds.

As we can see a large part of the FARC's demands represent the demands of peasant movements across Colombia (as illustrated by the discussion on the National Popular Agrarian Strike in Chapter 3). The FARC's and the Colombian government's expectations with regard to the length of the negotiations differ. President Santos apparently hopes to have completed the talks and the subsequent disarmament of the FARC in less than a year, but fulfilling at least some of the demands listed above cannot happen within such a short period. Moreover, the guerrilla organization would prefer to declare a ceasefire during this period instead of disarming, which is not what the government expects.

Even if the peace talks end successfully, such an outcome would not lead to an end to the armed conflict. There are several very significant reasons for this. The first is that Colombia's history has taught us so far that what is legislated for can be a world apart from what really happens on the ground. Between the 1930s and 1960s land reform laws were passed but their implementation was sluggish and failed to make a substantial change in the inequality of land ownership. Subsequently, such laws were reversed and the members of peasant movements were criminalized. Another illustration of the wide discrepancy between the law and the reality on the ground was the creation of paramilitary groups, which after being outlawed in 1989 was followed by a paramilitary boom in the 1990s. While the 1991 Constitution included the formal recognition of the rights of indigenous and Afro-Colombian communities, these people have been subjected to a relentless dispossession of their territories by local and foreign agroindustries and mining operations. While the creation of ZRCs was authorized by Law 160 of 1994, for the last 20 years the state has been oblivious to the peasants' need for the problem of landlessness to be addressed. The Victims and Land Restitution Law, which if carried out would constitute a small step in the direction in which the FARC want to move, is yet another example of the inconsistency between the law in theory and the practice in reality. Even if only two million hectares of land is recognized by the Santos administration as having been stolen (five times lower than the figure presented by numerous human rights organizations and social movements), and even if further displacements do not take place (which unfortunately is not the case), given the pace of the land restitution process it would take approximately 4,000 years to return all the land to the peasants who were forcibly expropriated.[4] In the previous chapter we already noted the numerous obstacles that make it highly unlikely that this will actually happen.

The second noteworthy factor hindering the possibility for peace with social justice is the fact that the Colombian state is a neoliberal state. As

Harvey explains, the latter always sides with 'a good business climate as opposed to either the collective rights (and quality of life) of labour or the capacity of the environment to regenerate itself' (2003: 70). Moreover, 'The coercive arm of the state is augmented to protect corporate interests and, if necessary, to repress dissent' (2003: 77). The freedom of private enterprises to operate, even when it is at the expense of the working poor and the environment, is considered beneficial for the country as a whole. The essence of the economic model promoted by President Santos stands in complete contrast to the radical reforms demanded by the FARC and by peasant, indigenous and Afro-Colombian movements across the country. The government's advancement of agribusiness and mining interests, combined with foreign investment and ownership of land, enshrined under the various FTAs, only aggravates the existing problems of landlessness, extreme asymmetries in land distribution, and lack of food autonomy, among many others. Not to mention the fact that even if the government would like to do otherwise, the freedoms that have been already given to private enterprises cannot be taken away peacefully, nor can the appropriation and monopolization of land and resources be reversed this way. None of the capitalist sectors would voluntarily give up their sources of capital. In any case, so far the government has not even expressed any such intention (and even if it made such promises it would be unlikely that these would ever materialize). Note that even in its rhetoric, the Santos administration speaks of land restitution (that is, returning stolen land) but not of redistributive land reform. The proposed Law of Land and Rural Development of 2012 (Ley de Tierras y Desarrollo Rural) makes this distinction clear. Even before the period of intense paramilitary-led forced displacement (1990–2006), Colombia's rural majority already had an acute need for land redistribution. Today, not surprisingly, the government only makes promises with respect to 'undoing' the damage caused by the illegal appropriation of land. However, the need to create a more egalitarian social structure has never been and will never be entertained by the state even in theory.

The third impediment to peace is the central feature of the present class system: the reliance on violence for capital accumulation. The operations of armed groups generate an ongoing dispossession which is being assisted by the paramilitary's multi-level networks inside major state institutions. The latter are ironically the very bodies that are supposed to implement land restitution and possibly land reform programmes (if such were agreed upon during the peace talks).

Whatever commitment has been made by the government to allocate land, its actual realization has been obstructed, as evidenced in the slaughter of those who have reclaimed their land or have organized to do so (as demonstrated in Chapter 4). In other words, while promises are being made and laws are being passed by the Colombian government, people are being killed in their efforts to receive what these promises and laws are supposed to deliver. Since the government continues to deny the present existence of paramilitary groups, thus erasing the historical nexus between the state, paramilitary forces and the capitalist classes, it is only natural to conclude that no change of the calibre necessary to break the engine of dispossession and repression is likely to happen. As a folk saying has it: 'You can wake up one who is sleeping but you can't wake up one who is pretending to be asleep.'

A final point: if hypothetically the government were to meet FARC's demands, it would have to channel a considerable amount of resources away from the military and into social programmes. It is quite naive to hope that this will happen, since even in a post-FARC era the state would always have a pretext, such as BACRIM or the existence of other guerrilla groups, to maintain its high level of militarization.

Third Principle: the Paramilitary and the State – A Dialectical Relationship

The relationship between the paramilitary and the state has always been dialectical in nature, meaning that it is a dynamic two-way relation in which each side shapes and affects the other's evolution. It is characterized by flows of weapons, intelligence, money and commodities of high monetary value such as illegal drugs, land and businesses. All of these sustain a wide spectrum of mutually beneficial activities, the success of which depends on the collaboration between members of each entity. The paramilitary's predatory violent activities generate the revenue, while the state provides the stamp of legitimacy, transforming the illegal into the legal (for example, the practices of INCODER discussed in Chapters 3 and 4). The paramilitary's terror strategies repress popular organizing in order to neutralize those who challenge capital and the status quo and who, according to the state, represent a threat to security, while state forces provide security back-up and reinforcement for paramilitary operations. The paramilitary ensures the disappearance of state enemies against whom the state has no concrete proof, while state employees ensure the disappearance of documents and witnesses that might serve as evidence of the paramilitary's human rights violations (Hristov 2009a).

This dialectical relationship is circular in nature. It began when the state laid the legal and military foundations for the existence of paramilitarism in the 1960s as it recruited and armed civilians to operate as paramilitary forces. This outward expansion from the centre (the state) towards sectors of civil society reached a new stage in the 1980s as the economically and politically dominant sectors of civil society began to set up paramilitary bodies themselves. The latter were outside the official state structure but developed in a continuous relation to it. The state tolerated them and provided military assistance in the form of weapons, training, bases, uniforms, transportation and so on. In the late 1990s, by the time of the unification of these groups under the name AUC, the paramilitary had achieved such a high degree of financial and military power and territorial control that it was able to establish mutually beneficial relationships with institutions beyond the state's coercive apparatus, such as the criminal justice system and the political system at all levels. This last development can be depicted as an inward movement where forces from outside the official boundaries of the state (the AUC and other paramilitary groups) penetrated state institutions. To conceptualize this dialectical relationship and the apparent permeability of the state, it is useful to refer to Jessop's understanding of the state, which 'does not exist as a fully constituted, internally coherent, organizationally pure and operationally closed system but is an emergent, contradictory, hybrid and relatively open system' (1990: 346).

Two processes can be identified as part of this circular dialectical relationship. The first (an outwardly directed one) has been the decentralization and outsourcing of violence such that the state no longer has a monopoly over the means of violence. This represents what I would like to call a socialization of the state's networks of terror, where the state's coercive apparatus expands outwards and incorporates civilians into its networks. Manifestations of this include: the initial creation of paramilitary groups under Plan Lazo in 1965; the creation of CONVIVIR in 1994; and the civilian informants and peasant soldiers of the Defence and Democratic Security Programme of 2002. The outcome of the socialization of the state's networks of terror has been units that have a civilian face but nonetheless engage in activities that serve the state's mission of security and the preservation of the status quo. Thus, while the means of violence are no longer solely in the hands of those officially employed in state institutions, they are used in accordance with the state's agenda to neutralize and control the 'internal enemy' (that is, guerrilla and social movements seeking to change the politico-economic model currently in place).

The process I have described here as an outward expansion of the state's coercive apparatus and the incorporation of civilians into its networks is accompanied by another kind of process (which can occur simultaneously) – the inward penetration of state institutions by paramilitary organizations. In other words, while the state's coercive apparatus extends into civil society by involving civilians in its security operations, the second process enables paramilitary power to exercise influence and control over the functioning of these institutions and the outcomes of their actions. It is then a question of controlling not only the means of violence but also most spaces inside the state. We may call this process the paramilitarization of the state or, alternatively, the institutionalization of paramilitarism. This, however, presents no danger to the state itself or the status quo since paramilitary power is in essence the power of the capitalist classes who use violence to advance their interests. Even the rest of the economically dominant classes who do not directly employ violence to dispossess or to suppress dissent and enrich themselves still benefit from the violence that suppresses the working majority, given the inherent conflict between the bourgeoisie and the labouring classes. The penetration of the state by paramilitary power is thus in reality the present-day expression of a long historical pattern in Latin American history – the state's lack of autonomy from the elite. Moreover, it is direct living evidence for Marx and Engels' claim in the *Communist Manifesto* that 'The executive of the modern state is but a committee for managing the common affairs of the whole bourgeoisie' (1848/1987: 23).

This dialectical relationship between the state and the paramilitary blurs the division between state and civil society and problematizes the scholarly treatment of these as two distinct areas of analysis. It also makes it more difficult to sustain the distinction between legal and illegal. The founding principle of the state of law – the equality of all before the law and the illegitimacy of any citizen using force to submit others to their interests – is destroyed. The existence of paramilitarism gives the state's coercive apparatus a peculiar quality – a kind of elasticity or flexibility. As Palacio put it:

STATE'S COERCIVE APPARATUS

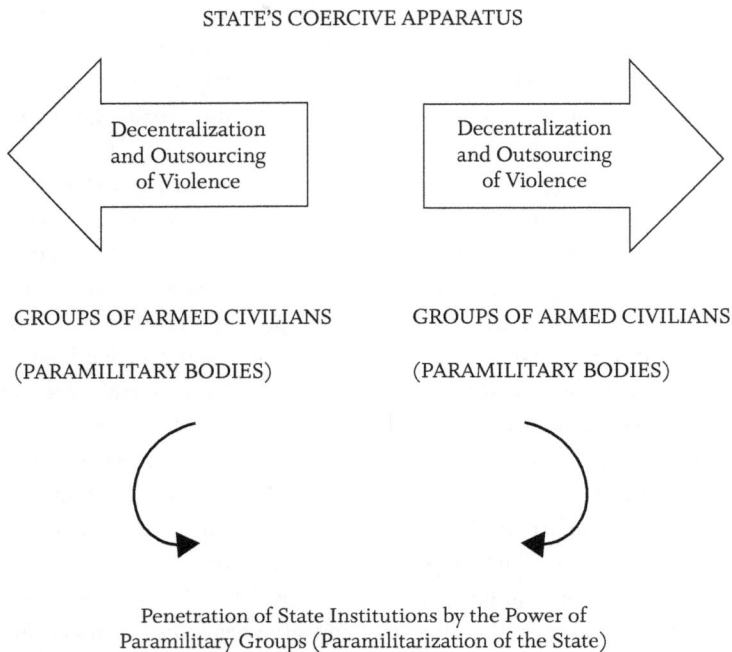

Decentralization
and Outsourcing
of Violence

Decentralization
and Outsourcing
of Violence

GROUPS OF ARMED CIVILIANS

(PARAMILITARY BODIES)

GROUPS OF ARMED CIVILIANS

(PARAMILITARY BODIES)

Penetration of State Institutions by the Power of
Paramilitary Groups (Paramilitarization of the State)

Figure 5.4   Outward and Inward Dynamics in the Dialectical Relationship
Between the State and the Paramilitary

to conceptualize parainstitutional mechanisms as separate from formal
institutional ones gives the mistaken impression that a real dividing
line exists between them. On the contrary, the Colombian State's
effectiveness lies precisely in its ability to combine seemingly opposed
structures and mechanisms. This is why we argue that in Colombia
parainstitutionality points to State flexibility. Colombia's particular
style of democracy requires such flexibility. (1991: 118–19)

The fluid boundary between state and civil society is also something that
has been emphasized by Jessop in his work on the state in capitalist society:

Above, around and below the core of the state are found institutions and
organizations whose relation to the core ensemble is uncertain. States
never achieve full closure or complete separation from society and their
precise boundaries are usually in doubt. Their operations also depend
on a wide range of micro-political practices dispersed throughout

society but concentrated and condensed in the core of the state. And they also enter into links with emergent state-like institutions above the nation-state at an inter-state level. (1990: 342)

Indeed, Colombia's landscape of violence illustrates this point through the multiple relations of interdependency among key violent actors (discussed earlier).

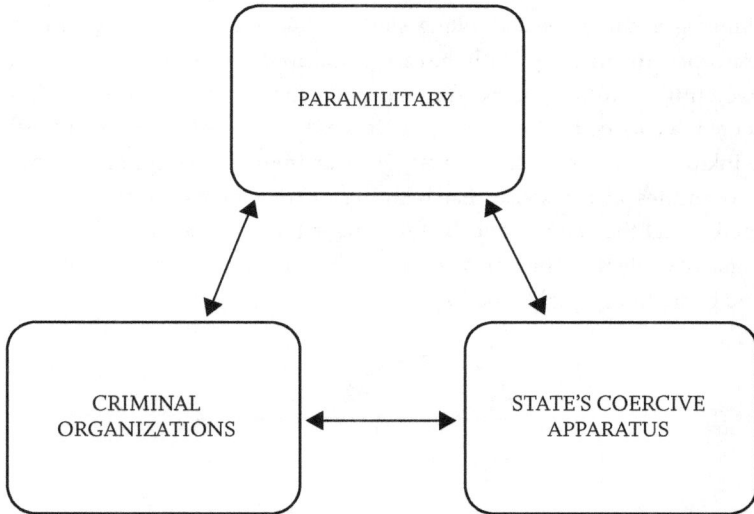

Figure 5.5   Interrelatedness of the State, Paramilitary Groups and Criminal Organizations

As mentioned earlier, paramilitary organizations own or control sectors of organized crime. The state has traditionally used paramilitary groups to carry out its 'dirty jobs'. At the same time, the state's coercive apparatus has in many ways assisted paramilitary illegal activities, including those linked to narco-trafficking. At this point it is important to realize that as the paramilitary's financial power has grown, these right-wing armed organizations have been increasingly providing lucrative work for state employees, especially those in the military and police, to supplement their salaries. Portraying the paramilitary mainly as an instrument at the disposition of the state and dismissing the fact that today the right-wing armed groups provide extra jobs for state employees, can obscure the paramilitarization of the Colombian state and the ways in which the state and the paramilitary complement each other. Figure 5.5 emphasizes the mutually determining characteristics of the relationships among the state, the paramilitary and criminal organizations. However, we must bear in mind that paramilitarism is a multidimensional phenomenon and is present inside most spaces within the state, not only the coercive apparatus. In sum, the paramilitary cannot be seen as a mere extension of the state's coercive apparatus nor as an entity totally independent of it. The two mutually constitute and interpenetrate each other. It is useful here to borrow Robinson's (2004) distinction between categorical thinking and relational thinking since it coincides with this idea of a dialectical relationship. With the categorical approach, Robinson explains, two entities are seen as existing independently from each other and, if there is a connection between them, it is external to both. With relational thinking on the other hand, we recognize an internal connection between two entities in the sense that both are distinct moments of the same totality and the one can only be fully understood in relation to the other. Figure 5.6 offers a depiction of what this dialectical logic between the state and paramilitary might look like.

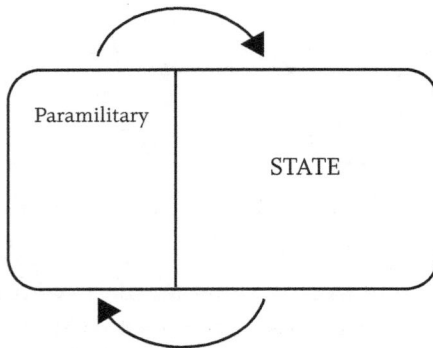

Figure 5.6   The Paramilitary and the State: Two Distinct Moments of the Same Totality

Paramilitarism and the Transnationalization of the Colombian State

One of the most exciting developments in the discipline of sociology today is the quest to develop a global perspective based on the realization of the magnitude and importance of the interconnectivity that exists above the level of any particular society. According to Burawoy (2011) the central focus of global sociology is global capitalism and its instantiations in different parts of the world. In my view, the most eloquent theory of global capitalism that is capable of accounting for the political counterpart of transnational capital has been developed by William Robinson. While approaches to the study of globalization which see the state as central to the reproduction of capitalist relations may recognize the global nature of the economic system, they still view the political system as being based on a nation-state-centred logic. According to Robinson (2004), some of these perspectives see nation-states as experiencing internationalization while others perceive transnational structures and institutions to be an extension of the nation-state. Robinson's theory of the TNS avoids either of these tendencies by making a crucial distinction between the state and the nation-state. In line with Marx's historical-materialist approach, the state as a coercive system of authority is understood as a congealment of a particular historically determined constellation of class forces and relations embodied in sets of political institutions. With the nation-state, the power relations and political institutions correspond to a specific territory. Robinson emphasizes the importance of not conflating the two. As discussed in Chapter 2, under capitalism the state is an instrument for securing the conditions of capitalist production. Under globalization, Robinson argues, the capitalist state has increasingly acquired the form of a TNS as the collective authority for the global ruling class. He defines the TNS is an emerging formation consisting of: 1) transformed nation-states, 2) supranational economic forums, and 3) supranational political forums. Nation-states are transformed and absorbed into the larger structure of the TNS through the various purposes they serve for transnational capital: border controls (that serve for population containment); transportation and communication infrastructure; availability of natural resources and cheapened labour for exploitation accomplished through the implementation of neoliberal policies; and provision of security.

Of particular interest to our discussion of paramilitarism in the context of the transnationalization of the state are the last two functions above. If by the transnationalization of the state we understand the process through which the nation-state becomes transformed and incorporated into the TNS, as the practices and ideologies of its institutions become increasingly tied to promoting the agenda of global capital, then I would argue that paramilitarism as a strategy is an integral part of the process of transnationalization. The substantiation of this argument is based on the implication of paramilitarism in the local/global capital-state configuration along several lines. First, it is necessary to review how the Colombian state has become part of the TNS. As Robinson (2004) has explained, the transnationalization of states and classes is interrelated and mutually determining. The Colombian case illustrates precisely this. Colombia has adopted neoliberal policies that have opened up the country's land, natural resources and labour to exploitation by transnational capital. At the same time, the capital-generating activities of a large segment of the Colombian elite (such as illegal drugs production, cash-crops cultivated by agribusinesses, and minerals extracted by mining companies) are part of global production chains. This is absolutely crucial because it reminds us that when the term 'global capital' is used in the context of Colombia, it does not necessarily imply the involvement of a foreign company. Robinson explains that the nation-state does not diminish in importance but rather tends to be influenced by social forces that are tied to the global economy. Such social forces in Colombia comprise both local and foreign enterprises. Therefore, the full beneficiaries of neoliberal restructuring in Colombia are not only foreign enterprises, but also all those local ones whose interests lie in global circuits of accumulation. Of course, there are specific elements of the neoliberal model that benefit all capitalists in Colombia (regardless of them being based on locally or globally oriented accumulation) by cheapening labour through a combination of primitive accumulation and labour deregulation. In addition to implementing neoliberalism, the Colombian state, through its coercive apparatus, has been the guarantor of global capital operating in Colombia.

Now that we have reviewed how the Colombian state has been transformed and incorporated into the TNS, we can bring paramilitarism into the equation. The phenomenon of paramilitarism is implicated in the process of transnationalization along four lines. First, it has facilitated the formation of a Colombian capitalist class linked to global circuits of accumulation. The principal social sectors funding paramilitarism – agribusinesses, extractive industries and drug-traffickers – are members of the TCC. They have a lifestyle characterized by the same trends in fashion, travel and entertainment as the elite from other parts of the world. They have luxury mansions with domestic servants, private security and expensive cars, their children study in elite schools, and they go to Europe and the US for vacations and shopping. Second, as demonstrated in the previous chapters, paramilitarism has been indispensable in the neoliberalization of the Colombian economy through primitive accumulation and the repression of opposition to neoliberalism's destructive impacts. Third, paramilitarism has enabled the Colombian state to guarantee security for foreign enterprises in the face of challenges by social movements and the guerrilla. Fourth, if the state is defined as a power relation embodied in a particular set of political institutions, and, as I demonstrated earlier, paramilitarism has nestled inside most state institutions, then paramilitarism is an ingredient of the political counterpart of global capital and, thus, is intertwined with the structure of the TNS itself.

In addition to being a strategy in the process of class and state trans-nationalization, the emergence, funding and functioning of paramilitary groups exhibits transnational linkages evident in the US military aid channelled through the state's coercive apparatus into the hands of private armed groups as well as in the presence of Israeli mercenaries who provided training during the 1980s and 1990s. Furthermore, as an ideology and a military project, paramilitarism is presently being exported, under the label of state-civil security partnerships, to its Latin American neighbours who have not overcome their own legacy of state terror. It is also being exported in the form of actual armed groups formed in Colombia and sent abroad to collaborate with local elites in Honduras and Venezuela, as discussed in Chapter 1. Clearly, paramilitarism is more than a form of parainstitutional violence. It is becoming an increasingly important strategy employed locally by agents of global systems of accumulation.

6

Conclusion

Violence has been a central and permanent feature of Colombia's history up to the present. This book has sought to contribute a new way of conceiving the social reality of Colombia's armed conflict by situating violence within the context of local and global processes of capital accumulation. In this concluding chapter, I briefly highlight the principal insights presented throughout the book, following which I discuss the broader inferences that can be drawn about the role of parainstitutional violence under capitalist social relations.

Chapter 2 considered the literature on the different modalities of organized private violence, where paramilitarism has been viewed as one such modality, and contended that although paramilitarism intersects with other forms of parainstitutional violence, it is qualitatively distinct. The chapter then assessed how useful existing conceptualizations are, not only for understanding the Colombian case but also for making broader inferences about the role of parainstitutional violence under capitalism. After examining three major approaches – the 'weak state', the paramilitary as a criminal actor, and the paramilitary as a subcontractor of state terror – I argued that none of them have been able to adequately situate paramilitary violence within unfolding processes of capital accumulation and economic globalization.

Chapters 3 and 4 revealed that capital formation and the participation of the majority of the population in capitalist production on terms that favour the dominant class (while reproducing its own marginalized position), are inextricably linked to the use of paramilitary violence. The historical review demonstrated that state and paramilitary violence has been a continuous and intrinsic element of the history of capitalism in Colombia that has made possible the progressive impoverishment of the working majority accompanied by the concentration of wealth and the establishment of control over resources and political power by a minority. As with most other Latin American societies, the destruction of people's sustainable livelihoods by dispossessing them from their means of subsistence, and the continuous transfer of control over land and other natural resources into fewer and fewer hands, have been stable features of Colombia's history in general. As Molano has rightly put it, the history of this country is a 'history of incessant, almost uninterrupted displacement' (2005: 81). Chapters 3 and 4 also traced the far-reaching consequences with regard to the massive forced displacement of people, concentration of landownership, consolidation of the presence of foreign enterprises, and repression of labour. These have been among the greatest challenges faced by the Left and those who struggle for social transformation in Colombia.

Chapter 4 challenged state discourses insinuating the obsolescence of paramilitarism and politically motivated violence by exposing the characteristics of the leaders and members of present-day non-guerrilla armed groups, the different forms of human rights violations carried out by them, and their mutually advantageous relationship with members of key state institutions. All of the empirical material presented showed that paramilitarism – as perhaps the most creative and intelligent effort by the state-elite enterprise to counteract revolutionary processes and at the same time allow for the use of violence in the acquisition of wealth – has not been eradicated with the so-called demobilization in 2006. To accept the claim that all non-guerrilla armed groups today are strictly tied to drug-trafficking and other forms of organized crime one must make oneself completely oblivious to the ongoing present-day forced displacement, threats and assassinations of labour unionists and members of other social movements, attacks against victims of past paramilitary crimes (or their families) who have stood up to demand justice and reparations, re-victimization of formerly displaced peasants who have attempted to recover their stolen land, right-wing threats against leaders of popular mobilizations, and the continuous collaboration between state officials and members of these armed groups.

Chapter 5 proposed an analytical framework for a social analysis of paramilitarism based on three principles that enable us to grasp the ways in which this multifaceted structure is the embodiment of the use of violence for the purpose of dispossession, the exploitation of resources and labour, and the suppression of dissent. These three principles can also be employed to determine whether a specific armed actor is paramilitary in nature or not, which can be of considerable significance given the myriad of parainstitutional actors and high levels of violence across Latin America today. Moreover, if we hope to eliminate the forces that engage in human rights violations, we first need to understand the conditions and mechanisms that sustain them and at the same time make their existence necessary. Each principle directs us to look for the social embeddedness of armed actors, including the economic and political linkages they exhibit.

The first and most important conclusion of this book concerns the relationship between violence and processes of capital accumulation. In order to elucidate the forces that have enabled the impoverishment, subordination and dependence of the greater part of the population within the current hierarchy of power, the analysis exposed the acts of violence that have dispossessed, disciplined and kept people inside exploitative and alienating social relations throughout the history of capitalist development in Colombia. Dispossession and labour repression have been the two kinds of violence that have secured the essential conditions for capital accumulation while giving rise to three major structural developments in Colombian society:

1) the formation of and/or sustainment of the principal capitalist classes – the landowners, agribusinesses, extractive industry enterprises, industrialists and drug-traffickers;
2) the progressive exacerbation of patterns of unequal wealth distribution; and
3) the implementation of neoliberal measures, such as the privatization of public enterprises and the elimination of restrictions on FDI, as well as the consolidation of free trade agreements with the US and Canada which further enhance foreign companies' access to Colombia's resources and labour force.

In sum, paramilitarism, employing both old and contemporary forms of dispossession, is part of the social meaning of capitalism.

The second conclusion that can be drawn concerns the ongoing nature of primitive accumulation as a geographically and historically pervasive element of capitalism which constantly seeks to recreate the basic prerequisites for its functioning – access to resources, labour force and markets. The above-mentioned two kinds of violence constitute methods of primitive accumulation that have been underway throughout the development of capitalism historically and have been especially accelerated as global capital penetrates deeper into the developing world through the exploitation of resources and people. The engagement of paramilitary groups in activities that lead to capital accumulation has not been recognized as capitalist by a considerable part of the academic literature due to the fact that activities such as drug-trafficking and forced displacement are seen as external to the sphere of the legal economy. They are regarded as phenomena peculiar to the armed conflict that have no consistent connection to the real course of capital accumulation. The underlying logic of such thinking echoes Weber's view of capitalism in which only the peaceful and rational acquisition of wealth is characterized as being capitalist in nature. The analytical framework I have proposed for conceptualizing paramilitarism helps us to re-think the social costs and mechanisms that sustain capitalism.

The third insight to be drawn from this study is that a formally democratic political regime can coexist with violence-based regulatory frameworks that often result in the perpetration of gross human rights violations. As Koonings and Krujit (2004) explain, this is precisely one of the noteworthy characteristics of contemporary Latin American societies – the de facto co-existence of formal constitutionalism and electoral democracy on one hand and the use of force to pursue economic or political interests on the other. The Colombian case illustrates precisely how this combination is made possible through the use of paramilitary force. Moreover, given the penetration of most state institutions by paramilitary power, it is imperative to view it not only as a military phenomenon but also as an institutional ensemble, operating both within and outside the state.

Finally, this work hopes to contribute to the theoretical approaches that seek to explain the transformation of the nation-state under neoliberalism and its gradual incorporation into the TNS (principally that of Robinson 2004), by emphasizing parainstitutional violence as an essential ingredient in this process. During the last 30 years processes of capitalist globalization have encouraged or supported the emergence of parainstitutional armed actors as a result of which the state no longer exercises a monopoly over the means of violence. Nonetheless, far from representing a threat to state power, these armed actors assist the state in performing its role of guaranteeing security for private capital (whether local or foreign). Therefore, the decentralization of violence must be understood not as a sign of state weakness, but rather as a more efficient strategy for securing foreign investment, protecting local landownership structures, enabling the further expropriation of small-scale property holders, and disempowering labour. The simultaneous connections between the paramilitary, the capitalist classes and the state are perhaps the best possible illustration of Marx's view of the capitalist state as being internally related to the social forces and structures that make up society rather than an independent sphere with its own logic (the way Weber saw it).

This work can serve as a departure point for subsequent investigations of paramilitarism. We have seen how the latter exacerbates class inequality. However, considering that class, gender and other forms of inequality mutually sustain each other, it would be useful to investigate how the phenomenon of paramilitarism contributes to or aggravates gender inequality in Colombia. Such a study could branch in several directions. One would be to examine how violence against women is used as a political tool to silence female activists, politicians and leaders of popular organizations and consequently to disempower social movements and impede systemic change. Another would be to consider how the cultural meanings that paramilitarism promotes encourage the further objectification and inferiorization of women.

This study also raises a more general question about capitalism and the well-being of humanity. Does capital inevitably violate human rights in order to exist? The Colombian experience challenges the Weberian understanding of capitalism as a non-violent enterprise. Given the systematic dismantling of the welfare state and deepening inequality even within developed countries, as well as the increasingly repressive and militaristic measures being taken against dissent, one might ask whether we are witnessing a pattern in which the violent tendencies of capitalism are becoming more and more pronounced in the countries of the Global North the way they already are in the Global South. Certainly in the case of Colombia, paramilitarism will continue to flourish for as long as the central feature of its politico-economic model is the existence of dominant classes whose aim is to maintain their power and privileges and enrich themselves by progressively dispossessing the working class and destroying all forces of resistance.

Notes

1. Brazil had 125 political killings during 15 years of military dictatorship, Bolivia had 243 in 17 years, Chile had 2,666 in 17 years, and Argentina had 9,000 in 8 years (Giraldo 1996).

2. Forced disappearance refers to a case where a person has been taken against his/her will by agents of the government or members of illegal armed groups and his/her whereabouts remain unknown.

3. Forced displacement refers to cases where people have been driven by means of violence or threats to abandon their place of residence out of fear for their own safety or that of their family.

4. A more comprehensive definition of parainstitutional violence is offered in Chapter 2.

5. The existing scholarly definitions of the term 'paramilitary' are explored in depth in Chapter 2.

6. The largest mass grave in Colombia was discovered in 2010 in the rural area La Macarena, Department of Meta, south of Bogotá. Approximately 2,000 bodies were buried there between 2002 and 2009 by the Colombian army. Military officials claimed those were the corpses of guerrilla fighters killed in combat; however the bodies were buried secretly without the requisite process of having the Colombian government certify that the deceased were indeed armed combatants as the army claimed (Kovalik 2010).

7. These are socialist-oriented revolutionary armed groups that seek to take over state power. The largest guerrilla organization in Colombia today is the Revolutionary Armed Forces of Colombia (Fuerzas Armadas Revolucionarias de Colombia, or FARC).

8. The so-called 'internal enemy' (a term used in US and Colombian counter-insurgency discourses) in reality has never included only the guerrilla, but has typically encompassed leaders (and frequently members) of non-armed social movements such as indigenous, peasants, women, students, labour unions, human rights activists, progressive educators and Leftist political parties.

9. In this book, the term 'social cleansing' refers to acts of violence directed against visibly poor members of society occupying public spaces, such as the homeless, panhandlers, street prostitutes, some informal vendors, and mentally ill individuals.

10. The HDI is a composite index measuring average achievement in three basic dimensions of human development: a long and healthy life, knowledge, and a decent standard of living (UNDP 2011).

11. Extreme poverty here is defined as cases where an individual lives on less than $47 a month (Prensa Latina 2012).

12. Measure of the deviation of the distribution of income (or consumption) among individuals or households within a country from a perfectly equal distribution. A value of 0 represents absolute equality, a value of 100 absolute inequality. For instance, the Gini coefficient for Canada is 32 (UNDP 2011).

13. It should be kept in mind that, as Briceno-Leon and Zubillaga (2002) point out, the homicide rate is a limited measure, as it does not take account of violence which does not result in death.

14. Land-grabbing leads to the privatization of water and the loss of food security and sustainable livelihoods for small-scale farmers, indigenous people, and independent fishermen. It also leads to the destruction of ecosystems, depletion of natural resources, and enhances unsustainable commodified models of agriculture.

15. Foreign investors are no longer only American or European but include companies from China, India, South Korea, Saudi Arabia and Qatar (Albinana 2012).

16. According to Mexican activists, the War on Drugs cost approximately 40,000 lives between 2006 and 2011, a period during which the murder rate rose 260 per cent (Bird 2011).

17. In Rio de Janeiro, para-state death squads are composed of off-duty, retired or suspended police officers, prison guards and firemen. In the state of Pernambuco, 70 per cent of all homicides in 2008 were attributed to death squads or so called extermination groups composed of agents of the state, particularly police (Pearce 2010).

18. These conferences are sponsored by the Northern Virginia-based contractor Continental Security and Interactive Solutions and are attended by presidents and mayors (Bird 2011).

19. Colombian armed forces are present in Honduras as part of the security cooperation agreement that Honduran President Porfirio Lobo signed with Colombia (Bird 2011).

20. The term 'narco-trafficking' is used throughout this book in accordance with Palacio's (1991) definition of it as the juridical and police transnationalization of the discourse around the trafficking of illegal drugs. The significance of his definition is that it highlights the function of the term as a political device used by governments, particularly the US, to justify repressive, disciplinary social control operations. I differentiate between 'narco-trafficking' and 'drug-trafficking'. I use the latter when I am referring to this activity as an illegal, internationalized business, which, as Palacio (1991) remarks, supports and contributes to capitalist accumulation.

21. Cosigo Resources Ltd, a Vancouver-based gold exploration company, was granted an exploration title in 2009 for the Yaigoje Apaporis, a one million

hectares national park which thanks to the initiative of the local indigenous population had been created for the purpose of conservation (Castro 2011).

22. Embridge, a Calgary-based energy corporation, is the biggest foreign investor in Colombia's largest pipeline, the Oleoducto Central SA (OCENSA) pipeline consortium. Thousands of peasants living along the route of the pipeline have been displaced as part of the security provisions (Gordon 2010).

23. While Conquistador Goldmines, a Canadian/Colombian subsidiary of the US-owned Corona Goldfields, was acquiring mining concessions in the Department of Bolívar in 1997, the AUC wreaked terror in the mining areas where artisan miners were operating, by burning villages, assassinating labour leaders and displacing 20,000 people (O'Connor and Bohorquez 2010).

24. Prime Minister Stephen Harper announced the Canadian government's intention to sign a free trade agreement with Colombia during his visit there in July 2007. The agreement was signed on 21 November 2008. The legislation was passed in the Canadian House of Commons on 14 June 2010 and the agreement came into effect on 15 August 2011.

25. The US-Colombia Trade Promotion Agreement was signed on 22 November 2006. The US Congress passed the agreement on 12 October 2011.

26. The local elite has played a central role in the development of capitalism in Colombia but has been understudied.

Chapter 2

1. Throughout the book I rely on the definition of capital accumulation as the central dynamic of capitalist development and a process whereby capital is expanded through the production, appropriation and realization of surplus value (Marshall 1998).

2. As I use the term 'class' throughout this chapter, I am referring to class under capitalism.

3. The relations of production are the relations of ownership, control and access that revolve around the means of production.

4. This work was published in 1849 in *Neue Rheinische Zeitung* and was later refined by Marx and appeared in 1859 as part of *A Contribution to the Critique of Political Economy*.

5. Necessary labour is the labour performed to produce the value necessary to ensure the reproduction of the labour-power (that is, the sustenance of the worker so he/she can stay alive). Surplus labour is the labour carried out after the necessary labour has been performed. 'The fact that half a day's labour is necessary to keep the worker alive during 24 hours does not in any way prevent him from working a whole day. Therefore, the value of labour-power and the value which that labour-power valorizes in the labour process are two entirely different magnitudes; and this difference is what the capitalist had in mind when he was purchasing the labour-power' (Marx 1867/1990: 300).

6. There are many other distinguished writers on the subject of capitalist globalization, such as Wallerstein (2000), Amin (2003) and Harvey (2003), to mention only a few. Robinson's (2004) work on capital–labour relations has been chosen here due to his eloquent account that simultaneously captures how the working and the capitalist classes have been affected by global forces during the last 30 years, and more specifically, his concept of the Transnational Capitalist Class (TCC).

7. In addition to the new features of class relations mentioned in the previous pages, a particularly interesting dimension is the rise of new dominant groups and capitalist factions tied to the global economy, or as the author calls them, the Transnational Capitalist Class (TCC). The latter is comprised of the owners and managers of transnational corporations (TNCs) and the private transnational financial institutions that drive the global economy. The TCC has developed a transnational class consciousness. 'In this sense it is a class-for-itself; whereas the global working class is a class-in-itself but not yet for-itself' (Robinson 2004: 31).

8. For other Marxist analyses of the state, see Miliband (1973) and Jessop (1990).

9. At the same time, the Colombian case represents a slight variation from one of the points Wood (1981) makes above. She theorizes capitalism as involving two formally separate but interconnected processes of coercion: 1) economic coercion accomplished by dispossession/primitive accumulation; and 2) extra-economic (political) coercion exercised by the state that sustains the conditions for the former. According to her, the moment of coercion appears as separate from the moment of appropriation. In Colombia, with the existence of paramilitary groups, the state and the appropriator are approximated further, since these organizations embody the fusion of the state and capitalists' initiatives to protect and advance processes of accumulation. Consequently, the two processes of economic and extra-economic coercion merge as capital engages in violence directly.

10. Under feudalism, land was not an absolute private property. The landlord had rights to it in return for military service. Peasants had use rights to it in exchange for services and agricultural produce.

11. 'The Parliamentary form of the robbery is that of "Bills for Inclosure of Commons", in other words decrees by which the landowners grant themselves the people's land as private property'; '3,511,770 acres of common land which between 1801 and 1831 were stolen from them and presented to the landlords by the landlords, through the agency of parliament...' (Marx 1867/1990: 885, 890).

12. Examples of such laws include the abolishment of the feudal tenure of land and the Bill for Enclosure of Commons.

13. For more such examples see Chapter 28, 'Bloody Legislation against the Expropriated', in *Capital*, Vol. I (Marx 1867/1990).

14. This is not meant to be an exhaustive review of this debate. For more on the latter, one of the best sources is *The Commoner* (www.thecommoner.org).

15. The World Health Organization defines violence in the following manner: 'The intentional use of physical force or power, threatened or actual, against oneself, another person, or against a group or community that either results in or has a high likelihood of resulting in injury, death, psychological harm, maldevelopment or deprivation' (Pearce 2010).

16. Since 2001 this kind of violence has also been denominated as 'terrorism' by US national security discourses and those of its Latin American allies.

17. Criminal violence can, of course, also be interpersonal for either material gain, or domestic (for example, spouse and child abuse).

18. Pereira and Davis (2000), Koonings and Krujit (2007), Mazzei (2009), Pearce (2010) and Sprague (2012) are leading in this field of research.

19. A crisis of hegemony occurs when political control can no longer be maintained through ideological manipulation, and therefore the dominant classes must resort to direct physical coercion by police and the armed forces (Burke 1999).

20. An in-depth discussion on the militarization of Colombia is offered in Chapter 3.

21. The transnational fractions of the capitalist classes in Latin America (found in agroindustries, finances, telecommunications, petrochemicals, banks and airlines) are the local contingents of a TCC whose economic interests reside in the global capitalist economy. The Latin American contingent of the TCC has developed conglomerates and private sector associations which have served as a key vehicle for organizing and exerting influence over state policies. Transnational elites seek to ensure the presence of cheap labour and access to natural resources and fertile land (Robinson 2004).

22. This is not meant to be an exhaustive review of paramilitarism worldwide.

23. In the case of Northern Ireland, this includes the political stand (in favour of or against British rule).

24. The main guerrilla group at that time was the Sendero Luminoso.

25. For more on this subject, see Degregori et al. (1996).

26. Granovsky-Larsen (forthcoming) explains that these forces do not always or consistently operate strictly as paramilitary groups, since they perform other non-paramilitary roles as well.

27. Examples of such groups, regarded by some as death squads, included the Anti-communist Liberation Armed Forces of Wars of Elimination (Fuerzas Armadas de Liberación Anti-comunista de Guerras de Eliminación, or FALANGE) and the White Warriors Union (Union de Guerreros Blancos, or UGB) (Global Security 2012).

28. Current research on paramilitary groups in other parts of Mexico is being carried out by Julie Mazzei.

29. A massacre is defined as the murder of four or more individuals at the same time and place by the same perpetrator.
30. Discussed in Chapter 3.
31. The Colombian state possesses military equipment (for example, various types of aircraft, aerial bombs, tanks, intelligence and radar electronic equipment) that is supplied by the US and is not available for purchase to any armed actor, such as the guerrilla.
32. For a review of the historical evolution in the meaning and usage of the term in the Colombian context, see Huhle (2001).
33. This author was cited above in relation to his concept of parainstitutionality and will be cited below in relation to his notion of the 'parainstitutional state'.
34. Hanson and Sigman (2011) summarize three main ways in which state capacity can be measured, according to theories of Economic Development; Stability and Change; and International Security. According to the first type of theory, state capacity refers to the state's ability to perform functions such as providing goods and services, protecting property rights, enforcing contracts, and maintaining order. Stability and Change theories define state capacity as the ability to maintain order through the repression of opposition, the accommodation of oppositional demands, and the legitimization of state power. According to International Security theories, state capacity has to do with intelligence and monitoring, control of borders, and authority over territories and social groups.
35. According to Tilly (2003), there are four possible combinations of levels of government capacity and levels of democracy and, thus, four types of regimes: high-capacity undemocratic; low-capacity undemocratic; high-capacity democratic; and low-capacity democratic.
36. For more on the relationship between forms of collective violence and regime types see Tilly (2003; 2005).
37. Examples of indicators of the state's coercive capacity, according to Hanson and Sigman (2011), include the ability to protect borders, authority presence over the entire territory, intelligence and monitoring, and presence of a professional military.
38. A more detailed discussion of paramilitarism and drug-trafficking appears in Chapter 5.
39. This book relies on the UN definition of human rights found in the UN Universal Declaration of Human Rights, https://www.un.org/en/documents/udhr

Chapter 3

1. The town of Santa Fe de Bogotá, today's capital city, was found in 1538.
2. The *hacienda* was a large landed estate (over 500 hectares), also known as *latifundio*, which first appeared during colonialism and continued for some

time after the formation of the independent nation-state. It resembled European feudal estates with regard to the following: 1) production consisted of many different agricultural crops and livestock and was oriented primarily towards ensuring self-sufficiency and secondarily towards sale at a local market; and 2) landlords leased a plot of land to peasants (*peones*, later on also called *terrajeros*) in exchange for compulsory labour-service for a certain number of days every week. By the mid 1800s many *haciendas* became capitalist farms as they became oriented towards commercial agriculture primarily for export and tenants became wage labourers employed on a contractual basis. Although the term '*hacienda*' is still used today in Colombia, it no longer carries the original meaning in terms of productive relations, but is rather used simply to accentuate the large size of the property and the colonial style architecture of the buildings within it.

3. The *encomienda* was a legal system that entailed forced labour through which the Spanish Crown granted a Spanish colonizer (*encomendero*) the right to forcibly extract a tribute in the form of labour, gold, or other products from a certain number of indigenous people from within a given geographic area. In exchange the colonizer was responsible for converting them to Catholicism and teaching them Spanish. The *encomendero* had property rights over the labour of the indigenous individuals but differed from slavery in the restrictions on inheritance, trading and relocation. The system was formally abolished in 1720, but had lost its effectiveness much earlier (Safford and Palacios 2002).

4. The *resguardos* were lands between 200 and 20,000 hectares in size awarded by the Spanish Crown to indigenous communities to be administered by an indigenous council. The most common form of land tenure in *resguardos* was collective, while there were some *resguardos* where the land was divided among individual families. After the end of colonial rule, most *resguardos* were expropriated and transformed into large rural estates privately owned by the non-indigenous elite (McGreevey 1971).

5. *Terrajeros*, also referred to as tenant farmers, were indigenous individuals who worked on *haciendas* in exchange for the right to have a home and a plot of land for subsistence on the territory of the *hacienda*.

6. The *minifundios* are mainly subsistence-oriented small farms less than five hectares in size (Safford and Palacios 2002).

7. The *palenque* was a settlement of escaped slaves and there is evidence that a high degree of communication, coordination and organization existed among *palenques* (Oquist 1980).

8. According to Marx's definition of primitive accumulation, the surplus extraction of gold and minerals by Spain also constituted a form of primitive accumulation by the metropolis.

9. After 1885, the 'Estados' became 'Departamentos' (Holmes, Gutierrez and Curtin 2008). Throughout this book the term 'Department' (as in 'Department

of Antioquia') is used as the equivalent of 'Departamento', when referring to geographic units.

10. All *resguardos* were subject to partition, and private ownership over these lands could be obtained through inheritance claims or purchase. Although theoretically all adults had equal rights to *resguardo* lands, in reality the apportioning of titles was subject to manipulation. There were cases when indigenous officers or leaders benefited from the division of their communal lands through establishing mutually beneficial alliances with neighbours in a position of power. Nevertheless, most of the gains were made not by indigenous inhabitants but by outsiders who were able to amass large land-holdings or multiple individual plots that totalled hundreds of hectares (Reinhardt 1988).

11. The civil wars took place in 1827–32, 1839–41, 1851–54, 1858–63, 1876–77, 1885, 1895 and 1899–1902 (Oquist 1980). The 1885 civil war made possible a radical change in the Constitution and in 1886 Colombia moved from a federal to a centralized republic. The constitution of 1886 lasted until 1991 (Safford and Palacios 2002).

12. The Liberal Party drew up its first programme in 1849 and the Conservative Party in 1849 (Pearce 1990).

13. This flow of foreign capital into private investment and public works during the 1920s was known as the 'Dance of the Millions' (Pearce 1990).

14. Some of the most powerful *gremios* that still exist today include the National Federation of Coffee Growers (Asociación Nacional de Cafeteros, or FEDERACAFE), the National Association of Industrialists (Asociación Nacional de Empresarios de Colombia, or ANDI), and the National Federation of Traders (Federación Nacional de Comerciantes, or FENALCO).

15. It has been estimated that by 2007 Colombian drug-traffickers owned 42 per cent of the best land in the country (Holmes, Gutierrez and Curtin 2008).

16. The PCC was founded in 1930 and called for the improvement of both rural and urban workers' rights and labour conditions. It supported rural militancy and by the 1950s had created a mass base with a significant peasant following (Brittain 2010).

17. The creation of ANUC was an attempt by the government to control the peasant movement in a non-repressive manner. For more on this subject see Rudqvist (1983).

18. In 1919 the government recognized the right to strike but also guaranteed the right of employers to hire workers to replace strikers (Pearce 1990).

19. For detailed accounts of La Violencia see Guillen (1963), Fals (1965), Torres (1970), Garcia (1971), Mesa (1972), Oquist (1980), Pearce (1990) and Safford and Palacios (2002).

20. La Violencia affected most intensely the coffee-production regions (particularly the Department of Cundinamarca, Tolima and Valle del Cauca)

as well as the *minifundia* areas in the Department of Boyacá, Santander and Norte de Santander (Pearce 1990).

21. The names of these communities were Tequendama, Viota, Genova, Sumapaz, El Davis, Suroeste del Tolima, 26 de Septiembre, Guayabero, Marquetalia, Rio Chiquito, El Pato and Agriari. They were spread across the Department of Tolima and parts of the Departments of Meta, Caldas, Cundinamarca, Cauca, Quindio, Risaralda and Huila (Brittain 2010).

22. The FARC was born with a revolutionary programme for equality of opportunities with equitable distribution of wealth, peace with social equality and sovereignty (Holmes, Gutierrez and Curtin 2008). This book does not deal with the many present-day complexities of the movement such as its internal structure, sources of funding, and controversial practices, since this is not the main subject matter. Therefore this historical overview of the FARC's emergence is not meant to provide a nuanced understanding of the movement's current dynamics and relationship to the rest of society. A discussion of the FARC's peace negotiations with the Colombian government in Havana initiated in 2012 is offered in Chapter 5.

23. The ELN was founded in 1964 by Colombian rebels inspired by the Cuban revolution and radical Catholic priests who were exponents of Liberation theology. Its relationship with the FARC had been characterized for many years by conflict. In 2013 ELN commander Nicolas Rodriguez announced that his organization had reached a peace agreement with the FARC. The latter in turn expressed its willingness to help in initiating peace talks between the government and the ELN. The government estimate of the ELN's size was 1,500 as of 2013 (Semana 2013).

24. The political party Patriotic Union (Union Patriotica, or UP) – the political arm of the FARC created as part of the peace negotiations held between the FARC and the government of President Belisario Betancur – had over 3,500 of its members murdered or disappeared by paramilitary groups between the mid 1980s and the early 1990s (Holmes, Gutierrez and Curtin 2008).

25. 'The Peace of the Rich is a War against the Poor' was a graffiti I saw on the wall of a building in Medellín.

26. The expression 'open veins' is borrowed from Eduardo Galeano's famous book *The Open Veins of Latin America*.

27. The term 'Washington Consensus' has come to symbolize the neoliberal agenda. The core of the neoliberal programme implemented in Latin American nations – austerity, trade liberalization, privatization and deregulation – was directly advocated by the US Institute for International Economics (IIE) in its publication *Toward Renewed Economic Growth in Latin America* (1986). In 1989 the Institute convened a conference to explore the extent to which these measures were indeed undertaken in the region. The participants, including representatives from international financial institutions, economic agencies of the US government, and the US Federal Reserve Board, concluded the

conference by recommending policies that were still needed in Latin America. John Williamson, a senior official at the IIE, made a list of ten such reforms under the name 'Washington Consensus' consisting of: fiscal restraint, public expenditures in fields with high economic returns, tax reform, financial liberalization, competitive exchange rate, trade liberalization, removal of barriers to foreign investment, privatization of state-owned enterprises, deregulation, and security for property rights (Williamson and Kuczynski 2003).

28. Before 2001, Colombian citizens could legally mine on public lands (O'Connor and Bohorquez 2010).

29. The privatization of banks leaves small-scale farmers without access to credit and other critical resources which is detrimental to their survival (Lefeber 2003).

30. Statistics on assassinated unionists from the late 1980s until the present vary according to the organization providing the data. This is mainly due to the fact that in the late 1980s and early 1990s records of attacks against unionists were not always compiled. Only in the second half of the 1990s did the ENS as well as international organizations such as ICTUR and ICFTU begin producing yearly accounts of the murder of unionists in Colombia. At the low end of the estimate, according to US Labour Education in the Americas Project (USLEAP 2011), between 1986 and 2010 over 2,800 unionists were killed.

31. For more on the Defense and Democratic Security Program see Hristov (2009b).

32. For 58 years DAS was considered the state's central and most powerful intelligence institution. Among its functionaries were bodyguards, criminal investigators and customs officers. On 31 October 2011, President Santos signed Decree 4057 which put an end to this institution after numerous scandals around human and civil rights violations in which it had been involved. The new intelligence agency that is solely in charge of state security is the National Colombian Agency of Intelligence (Agencia Nacional de Inteligencia Colombiana, or ANIC), which is regulated by the Ministry of Defence. Former employees of DAS were transferred to ANIC, the Ministry of Interior, the Ministry of Foreign Relations, the National Police and the Fiscalia (El Tiempo 2011). ANIC began operating in 2012 and its director is Alvaro Echandia, former commander of the navy.

33. Still, Colombia's own dedication of resources to militarization should not be underestimated.

34. According to the Colombian government, Plan Colombia's components included: the fight against narco-trafficking and terrorism; strengthening of law and justice and promotion of human rights; opening of markets; comprehensive social development; comprehensive assistance for the displaced population; and demobilization, disarmament and reintegration (PDA 2007).

35. For a detailed account of the development of paramilitary groups see Hristov (2009b).

36. The initial nucleus of the AUC was the Peasant Self-defense of Córdoba and Urabá (Autodefensas Campesinas de Córdoba y Urabá, or ACCU) along with seven other groups: Autodefensas de los Llanos Orientales, Autodefensas de Ramon Isaza, Autodefensas del Cesar, Autodefensas del Magdalena Medio, Autodefensas de Santander y Sur del Cesar, Autodefensas del Casanare, Autodefensas de Cundinamarca and Autodefensas de Puerto Boyacá.

37. The following definitions are grounded in the context of the Colombian armed conflict and may or may not be fully applicable to other countries.

38. There are cases where the goal of forced displacement is not dispossession (that is, the goal is not to appropriate the property of the person who abandons it due to threats or violence). Such are usually the cases in urban areas, where the paramilitary selects certain individuals (for example, a union leader, professor, judge, or politician) who due to the nature of their work or activism represent a threat to the state, paramilitary groups or capitalist enterprise. Under such circumstances the individual (and their family) is forced to leave their home and migrate to another city or emigrate abroad. Such occurrences do not necessarily lead to dispossession and take place on an individual level (in contrast to displacement in rural areas which most often affects an entire community or area).

39. After 2003 INCORA became the Colombian Institute for Rural Development (Instituto Colombiano de Desarollo Rural, or INCODER). Both INCORA and INCODER have been equally involved in legalizing the illegal appropriation of land by the paramilitary.

40. The Cattle-ranchers' Fund (Fondo de Ganaderos) has been the biggest buyer of such land.

41. A *testaferro* is a person who is paid to have a property ascribed to their name (a front man).

42. The Colombian state's estimate is seven million (Comisión Nacional de Reparación y Reconciliación 2009).

43. Former President Alvaro Uribe's ties to the paramilitary are numerous, and often of a direct and personal nature. The following are only a couple of examples. His estate Guacharacas in San Roque, Department of Cesar, has been the headquarters for a paramilitary group of 40 men led by alias Beto. Uribe's brother Santiago also has *haciendas* that have served as bases for the paramilitary groups (CINEP 2005b). For his 2002 election campaign, Uribe received $40,000 from Enilce Lopez who was a manager of paramilitary businesses at the time (Moreno 2006). For more on Uribe's paramilitary background see Contreras (2002).

44. The Congress consists of the Senate (*Senado*) with 102 seats and the Chamber of Representatives (Cámara de Representantes) with 166 seats.

45. For extensive examples illustrating each of these connections between the paramilitary and the state military, see Hristov (2009b).

46. Particularly the Popular Feminist Organization (Organización Femenina Popular, or OFP).

47. The implementation of this project would cost the state 576 billion pesos along with the 2 million pesos compensation per family. ZRC in the Department of Norte de Santander would require approximately 925,000 acres.

48. They are found in Calamar (Department of Guaviare), Cabrera (Department of Cundinamarca), El Pato (Department of Caquetá), el sur de Bolívar (Department of Bolívar), el valle del Río Cimitarra (Department of Antioquia and Department of Bolívar) and Cuembí and Comandante (Department of Putumayo) (Celis 2013).

49. A large part of these form part of the demands presented by the FARC during the peace talks with the Colombian government in Havana which are discussed in Chapter 5.

Chapter 4

1. By using the term 'illegal armed groups' I refer to non-guerrilla armed groups present in the period 2006–14. The Colombian government has labelled these organizations as BACRIM while I argue that they are paramilitary in nature.

2. Of course in the case of Colombia, unlike Central America, the existence of the guerrilla continues to be the main element of the state's security ideology.

3. The loss of property rights if the land is deemed to have been abandoned applies in the following cases: 1) where peasants were allocated grants of small land-holdings by the government between the 1960s and the 1990s and had to make re-payments every month; and 2) in cases where peasants occupied a *baldío* and farmed it continuously for at least ten years, after which they were granted a land title.

4. See *La Economia de los Paramilitares* by Corporacion Nuevo Arco Iris (2011), Bogotá.

5. This is not an exhaustive list.

6. This is an estimate by MOVICE. The Colombian state's estimate is seven million (Comisión Nacional de Reparación y Reconciliación 2009).

7. For more on this law and its decrees, see the government's website http://www.leydevictimas.gov.co

Chapter 5

1. I do not consider the paramilitary to be a political movement since they are a political power already inside the state.

2. According to the Canadian Encyclopedia (2012), 'organized crime' can be defined as 'two or more persons consorting together on a continuing basis to participate in illegal activities, either directly or indirectly, for gain'. The sources of organized crime revenues could be broken down as follows: pornography, prostitution, bookmaking, gaming houses, illegal lotteries and other gambling, loansharking and extortion. Other activities considered to constitute organized crime are: white-collar crime (for example, insurance and construction frauds and illegal bankruptcies), arson, bank robberies, motor vehicle thefts, computer crimes and counterfeit credit cards and illegal drugs.

3. It should be noted that among the founders of paramilitary organizations there were some, such as alias Doble Cero, who were strongly opposed to any close relationship between paramilitary bodies and drug-traffickers, leading to some internal disagreements and the eventual separation from the AUC.

4. By early 2013 there were 16 court rulings ordering the return of illegally appropriated land totaling 500 hectares (HRW 2013).

References

Acevedo, B. (2008, September). 'Ten years of Plan Colombia: an analytic assessment'. The Beckley Foundation Drug Policy Programme.

ACIN (Asociación de Cabildos Indígenas del Norte del Cauca). (2005, December 15). 'Declaración sobre el tratado de libre comercio entre Colombia y Estados Unidos, TLC'. http://www.nasaacin.org

Agencia de Prensa. (2010, June 8). 'Por no pagar sus deudas, bancos se quedarían con tierras de desplazados'. http://www.nasaacin.org/noticias.shtml?x=11743

Agencia Prensa Rural. (2005, July 17). 'Paramilitaries kill Campesino'. *Agencia Prensa Rural*. http://www.prensarural.org/wnu20050717.htm

— (2013, October 1). '19 de agosto: Paro Nacional Agrario y Popular'. http://prensarural.org/spip/spip.php?article11620

— (2014, January 22). 'La paz sacrificando al movimiento social'. http://prensarural.org/spip/spip.php?article13174

AI (Amnesty International). (2004, October 13). 'Colombia: scarred bodies, hidden crimes – sexual violence against women in the armed conflict'. http://web.amnesty.org/library/print/ENGAMR230402004?open&of=ENG-Col

— (2005a, September 7). 'Urgent action: fear for safety/disappearance: Colombia'. Index no. AMR 23/029/2005. http://web.amnesty.org/library/Index/ENGAMR230292005?open&of=ENG-351

— (2005b, July 13). 'Colombia: President Uribe must not ratify impunity law'. Index no. AMR 23/021/2005. http://web.amnesty.org/library/Index/ ENGAMR230212005

— (2006a, May 10). 'Fear for safety of journalist'. Online Documentation Archive, 130/06. http://web.amnesty.org/library/Index/ENGAMR230212006?open&of=ENG-346

— (2006b, May 12). 'Urgent action: fear for safety of human rights lawyers, Colombia'. Online Documentation Archive. http://web.amnesty.org/library/ Index/ENGAMR230142005?open&of=ENG-346

— (2007, July). 'Colombia killings, arbitrary detentions, and death threats – the reality of trade unionism in Colombia'. http://www.amnesty.org/en/library/asset/AMR23/001/2007/en/b6801d38-d3ba-11dd-a329-2f46302a8cc6/ amr230012007en.pdf

— (2012). 'The victims and land restitution law: an Amnesty International analysis'. http://www.amnesty.org/en/library/info/AMR23/018/2012/en

Albinana, A. (2012, January 2). 'Luchas de paises ricos por controlar tierras fértiles en el mundo'. *El Tiempo*. http://www.eltiempo.com/mundo/africa/lucha-de-paises-ricos-por-controlar-tierras-fertiles-en-el-mundo_10932992-4

References 185

American Heritage Dictionary of the English Language. (2009). 'Paramilitary'. Boston and New York: Houghton Mifflin Company. http://ahdictionary.com/word/search.html?q=paramilitary

Amin, S. (2003, October). 'World poverty, pauperization and capital accumulation'. http://monthlyreview.org/2003/10/01/world-poverty-pauperization-capital-accumulation

Appelbaum, N. P. (2003). *Muddied Waters: Race, Region, and Local History in Colombia, 1846–1948*. Durham, NC: Duke University Press.

Araujo, M. C. (2011, February 26). 'Que pasó con la ex-ministra de relaciónes exteriores?' http://www.semana.com/noticias-enfoque/maria-consuelo-araujo/152514.aspx

Arias, E. D. (2006). *Drugs and Democracy in Rio de Janeiro: Trafficking, Social Networks, and Public Security*. Chapel Hill and London: University of North Carolina Press.

Arnson, C. J. (2000). 'Window on the past: a declassified history of death squads in El Salvador'. In B. B. Campbell and A. D. Brenner (eds.), *Death Squads in Global Perspective: Murder with Deniability*. New York: St. Martin's Press, pp. 85–124.

Aviles, W. (2006). 'Paramilitarism and Colombia's low-intensity democracy'. *Journal of Latin American Studies* 38, pp. 379–408.

Bannerji, H. (2001). 'Writing India, doing ideology'. In *Inventing Subjects: Studies in Hegemony, Patriarchy and Colonialism*. New Delhi: Tulika Books, pp. 18–53.

— (2003). 'The tradition of sociology and the sociology of tradition'. *Qualitative Studies in Education* 16(2), pp. 162–3.

Bastos, S. (2004). *Etnicidad y fuerzas armadas en Guatemala*. Guatemala: FLACSO.

BBC (British Broadcasting Corporation). (2011, May 25). 'Colombia: law for victims passed by Senate'. http://www.bbc. co.uk/news/world-latin-america-13542244

— (2011, May 29). 'Colombia details scale of land stolen in civil conflict'. http://www.bbc.co.uk/news/world-latin-america-13591860

Beale, M. (2011, December 28). 'The CARICOM blueprint for illicit drug trafficking'. Council on Hemispheric Affairs. http://www.coha.org/the-caricom-blueprint-for-illicit-drug-trafficking

Bejarano, A. M. and Pizarro, E. (2002, April). 'From "restricted" to "besieged": the changing nature of the limits to democracy in Colombia'. University of Notre Dame, Helen Kellog Institute for International Studies. Working Paper no. 296.

Berglund, S. (1982). *Resisting Poverty – Perspectives on Participation and Social Development: The Case of CRIC and the Eastern Rural Region of Cauca in Colombia*. Umea, Sweden: University Umea.

Bergquist, C. W., Peñaranda, R. and Sánchez, G. (eds.) (2001). *Violence in Colombia, 1990–2000: Waging War and Negotiating Peace*. Wilmington, DE: Scholarly Resources.

Bird, A. (2011, December 8). 'Feeding the monster: militarization and privatized "security" in Central America'. http://www.upsidedownworld.org/ main/international-archives-60/3347-feeding-the-monster-militarization-and-privatized-security-in-central-america

Boletín Virtual del Informe para el Exámen Periódico Universal. (2009, June 5). 'Impunidad y vulneración de los derechos de las víctimas a la verdad, la justicia y la reparación. No. 5'. http://ddhhcolombia.org.co/files/Boletin_N1/ Boletin_EPU5. html

Bond, P. (2005, October 27). 'Dispossessing Africa's wealth'. *Pambazuka News.* http:// www.pambazuka.org/en/category//30074

Bonefeld, W. (2001). 'The permanence of primitive accumulation: commodity fetishism and social constitution'. *The Commoner.* http://www.thecommoner.org

Bonilla, T. C. (2013, September 9). 'Que esta sucediendo en Colombia? El Paro Agrario Popular'. *Voces Semanario.* http://voces.org.sv/2013/09/09/que-esta-sucediendo-en-colombia-a-proposito-del-paro-agrario-y-popular

Briceno-Leon, R. and Zubillaga, V. (2002). 'Violence and globalization in Latin America'. *Current Sociology* 50(1), pp. 19–57.

Brittain, J. (2010). *Revolutionary Social Change in Colombia: The Origin and Direction of the FARC-EP.* London: Pluto Press.

Brockett, C. D. (1990). *Land, Power, and Poverty: Agrarian Transformation and Political Conflict in Central America.* Boston: Unwin Hyman.

Burke, B. (1999). 'Antonio Gramsci and informal education'. *Encyclopedia of Informal Education.* http://www.infed.org/thinkers/et-gram.htm

Burawoy, M. (2011). 'Global Sociology, Live'. Lecture for Sociology 185 at the University of Berkeley, California. http://burawoy.berkeley.edu/syllabus/185.pdf

Cabrios, A.S. (2014, January 29). 'Santos more optimistic than a year ago on Colombia peace talks'. Deutsche Presse-Agentur GmbH. http://latinotimes.com/ latinos/1126019-santos-more-optimistic-than-a-year-ago-on-colombian-peace-talks.html

Calvo, H. (2004, December 30). 'Privatized conflict'. ZNet. http://www.zmag.org content/showarticle.cfm? SectionID=9&ItemID=6939

Cambio. (2007, March 30). 'Regreso forzoso'. http://www.cambio.com.co/ especialescambio/home/ARTICULO-PRINTER_FRIENDLY-PRINTER_FRIENDLY_CAMBIO-3499788.html

Cambridge Advanced Learner's Dictionary and Thesaurus. (2012). 'Paramilitary'. Cambridge: Cambridge University Press. http://dictionary.cambridge.org/ dictionary/british/paramilitary_1?q=Paramilitary

Campbell, B. B. (2000). 'Death squads: definition, problems, and historical context'. In B. B. Campbell and A. D. Brenner (eds.), *Death Squads in Global Perspective: Murder with Deniability.* New York: St. Martin's Press, pp. 1–26.

— and Brenner, A. D. (eds.). (2000). *Death Squads in Global Perspective: Murder with Deniability.* New York: St. Martin's Press.

Canada. Prime Minister. (2009, June 10). 'PM meets with Colombian President Álvaro Uribe'. http://pm.gc.ca/eng/media.asp?id=2629

Canadian Encyclopedia. (2012). 'Organized crime'. http://www. thecanadian encyclopedia.com/articles/organized-crime

Caracol (2013, May 1). 'Condicion Laboral en Colombia segun la Escuela Nacional Sindical'. http://www.caracol.com.co/noticias/actualidad/condicion-laboral-en-colombia-segun-la-escuela-nacional-sindical/20130501/nota/1890985.aspx

Carpenter, T. G. (2005, November 15). 'Mexico is becoming the next Colombia'. *Foreign Policy Briefing* No. 87.

Castro, A. (2002, July 29). 'US military bases in Colombia'. *ANNCOL*. http://www.anncol.com

Castro, C. (2011, May 21). 'El "Avatar" Colombiano'. No. 1516. http://www.semana.com/nacion/avatar-colombiano/157077-3.aspx

Castro, Y. (2012, June 17). 'Restitución de tierra en la administración Santos: ¿éxito o fracaso?'. http://razonpublica.com/index.php/conflicto-drogas-y-paz-temas-30/3040-restitucion-de-tierra-en-la-administracion-santos-iexito-o-fracaso.html

Celis, L. E. (2013, March 13). 'La embarrada del ministro de Agricultura'. *Agencia Prensa Rural*. http://prensarural.org/spip/spip.php?article10439

Chan, S. (1999). 'The warlord and global order'. In P. Rich (ed.), *Warlords in International Relations*. London: Macmillan Press Ltd, pp. 164–72.

Chomsky, N. (2000, June). 'The Colombia Plan: April 2000'. *Z Magazine*. http://www.chomsky.info/articles/200006.htm

— (2003). *Hegemony or Survival*. New York: Metropolitan Books.

Christian Peacemakers Team. (2006). 'Informe de derechos humanos 2006'. http://www.cpt.org

CINEP (Centro de Investigación y Educación Popular). (2002). 'Deuda con la humanidad 2002: Santa Fe de Ralito y la legitimación definitivo del paramilitarismo'. *Noche y niebla*. http://www.nocheyniebla.org/ casotipo/ deuda/2002.pdf

— (2005a, September 26). 'Casos registrados entre enero y junio 2005'. *Noche y niebla* 31, 84. http://www.nocheyniebla.org/files/u1/31/pdf/05casos31.pdf

— (2005b). 'Deuda con la humanidad – El General Rito Alejo del Río, baluarte del paramilitarismo bajo el blindaje de la impunidad'. *Casos Tipo*. http://www.nocheyniebla.org/files/u1/casotipo/deuda/html/pdf/deuda15.pdf

— (2005c). 'Deuda con la humanidad – Estado y paramilitares: vinculos inocultables; revelaciones de parte y parte'. *Casos Tipo*. http://www.nocheyniebla.org/ files/u1/ casotipo/deuda/html/pdf/deuda14.pdf

— (2005d). 'Deuda con la humanidad – Los gobiernos de los Estados Unidos y el paramilitarismo colombiano'. *Casos Tipo*. http://www.nocheyniebla.org/ files/u1/ casotipo/deuda/html/pdf/deuda19.pdf

— (2005e, March). 'Violaciones a los derechos humanos semestre julio-diciembre 2004'. *Noche y niebla* 30, 277. http://www.nocheyniebla.org/files/u1/30/ pdf/16Diciembre20.pdf

— (2007, June 7). 'Buenaventura riqueza, genocidio y hambre'. Centro de Medios. http://colombia.indymedia.org/news/2007/06/67228.php

CIP (Center for International Policy). (2003, September 2). 'Just the facts: a civilian guide to US defense and security assistance to Latin America and the Caribbean

by the Latin America Working Group Education Fund'. http://www.ciponline.
org/facts/co.htm

CIP – Colombia Program. (2008, January 10). 'Osorio devastated the Fiscalia'. http://
www.cipcol.org/?p=521

CJL (Juridic Corporation Liberty – Corporación Jurídica Libertad). (2010, June 2).
'La Universidad de Antioquia hace la vista ciega al paramilitarismo que desde
hace rato penetro la institución'. http://www.cjlibertad.org/index.php?
option=com_content&view=article&id=331:la-universidad-de-antioquia-hace-
la-vista-ciega-al-paramilitarismo-que-desde-hace-rato-penetro-la-institucion&c
atid=29:pronunciamientos&Itemid=27

CODHES (Consultoria para los Derechos Humanos y el Desplazamiento). (2009,
April 22). 'Víctimas emergentes: desplazamiento, derechos humanos, y conflicto
armado en 2008'. Boletín Informativo de la Consultoria para los Derechos Humanos
y el Desplazamiento 75. http://www.codhes.org

— (2012, March). 'Desplazamiento creciente y crisis humanitaria invisibilizada'.
Boletín para la Consultoria para los Derechos Humanos y el Desplazamiento 79.
http://www.codhes.org/images/stories/pdf/bolet%C3%ADn%2079%20
desplazamiento%20creciente%20y%20crisis%20humanitaria%20visible.txt.pdf

Collier, P. (2006). Economic Causes of Civil conflict and their Implications for Policy.
Washington, DC: World Bank Development Research Group.

— and Hoeffler, A. (1998). 'On economic causes of civil war'. Oxford Economic Papers
50(4) (1998), pp. 563–73.

Collins, R. (1986). Weberian Sociological Theory. Cambridge: Cambridge University
Press.

Collins English Dictionary. (2003). 'Paramilitary'. New York: HarperCollins Publishers.
http://www.thefreedictionary.com/paramilitary

Colombia Week. (2004, September 27). 'Hitmen murder sociologist on Atlantic
Coast'. http://groups.yahoo.com/group/NativeNews-SOUTH/message/151

Comisión Nacional de Reparación y Reconciliación. (2009). 'El despojo de tierras
y territorios: Aproximación conceptual'. http://www.memoriahistorica-cnrr.org.
co/s-informes/informe-12

Contreras, J. (2002). Biografía no autorizada de Álvaro Uribe Vélez. Bogotá: Editorial
La Oveja Negra.

Corporacion Nuevo Arco Iris. (2011). La Economia de los Paramilitares. Bogotá.

Corriente Marxista. (2009, May 24). 'La nueva etapa de la lucha de clases y las
perspectivas para la construcción de una … Por corriente marxista internacional-
colombia'.http://colombia.elmilitante.org/colombia/pda/24-la-nueva-etapa-de-
la-lucha-de-clases-y-las-perspectivas-para-la-construccion-de-una-direccion-
revolucionaria.html

Council on Foreign Relations. (2005, November). 'Northern Ireland Loyalist
paramilitaries'. http://www.cfr.org/terrorist-organizations/northern-ireland-
loyalist-paramilitaries-uk-extremists/p9274

Craig-Best, L. and Shingler, R. (2003). 'The Alto Naya massacre: another paramilitary outrage'. *Colombia Journal*. http://www.colombiajournal.org

Crandall, R. (1999). 'The end of civil conflict in Colombia: the military, paramilitaries, and a new role for the United States'. *SAIS Review* 19(1), pp. 223–37.

— (2002). *Driven by Drugs: US Policy Toward Colombia*. Boulder: Lynne Rienner.

CSN (Colombia Support Network). (2005, September 7). 'Forced recruitment of children by demobilized paramilitaries in Medellín'. http://www. colombiasupport. net/news/2005/09/forced-recruitment-of-children-in_07.html

— (2007, October 8). 'Son of Coca-Cola worker kidnapped and tortured'. http://www.colombiasupport.net/news/2007/10/son-of-coca-cola-worker-kidnapped-and.html

Cubides, F. (2005). 'Narcotrafico y Paramilitarismo: Matrimonio Indisoluble?' In A. Rangel (ed.), *El Poder Paramilitar*. Bogotá: Planeta, pp. 205–53.

Dalla Costa, M. (2004, Autumn/Winter). 'Capitalism and reproduction'. *The Commoner*. http://www.thecommoner.org

DANE (Departamento Nacional Administrativo de Estadisticas). (2009, August 24). 'Misión para el empalme de las series de empleo, pobreza y desigualdad'. http://www.buenastareas.com/ensayos/Mision-Para-El-Empalme-De-Las/185892.html

De Angelis, M. (2001, September). 'Marx and primitive accumulation: the continuous character of capital's "enclosures"'. *The Commoner*. http://www.thecommoner. org

Degregori, C. I., Ponciano del Pino, J. C. and Orin, S. (1996). *Las Rondas Campesinas y la derrota de Sendero Luminoso*. Lima: Instituto de Estudios Peruanos.

Diamond, L., Hartlyn, J. and Linz, J. (1999). 'Introduction: politics, society, and democracy in Latin America'. In L. Diamond, et al. (eds.), *Democracy in Developing Countries: Latin America*. Boulder, CO: Lynne Rienner Publishers, pp. 1–19.

Domingo, J. (2011, November 5). 'Reducing inequality should be a political priority?'. IPS. http://ipsnews.net/news.asp?idnews=105742

Donahue, S. (2005, April 16). 'Palm growers and paramilitaries in Urabá'. *Narcosphere*. http://narcosphere.narconews.com/story/2005/4/16/181218/755

Dudley, S. (2004). *Walking Ghosts: Murder and Guerrilla Politics in Colombia*. New York: Routledge.

Duncan, G. (2006). *Los Señores de la Guerra: De Paramilitares, Mafiosos y Autodefensas en Colombia*. Bogotá: Editorial Planeta Colombiana.

Duncan, J. (2013, November 6). 'The auditor and the hitmen'. http://www.sacsis. org.za/site/article/1833

Duque, H. (2013, July 15). 'Catatumbo: Paro Nacional Agrario y Solidario'. *Telesur*. http://www.telesurtv.net/articulos/2013/07/15/catatumbo-paro-nacional-agrario-y-solidario-4220.html

El Colombiano (2010, September 24). 'Farc quedan heridas de muerte sin el "Mono Jojoy"'. http://www.elcolombiano.com/BancoConocimiento/H/historico_cayo_mono_jojoy_lcg_24092010/historico_cayo_mono_jojoy_lcg_24092010.asp

El Comercio. (2012, January 19). 'Al menos 29 sindicalistas fueron asesinados en 2011 en Colombia'. http://www.elcomercio.com/mundo/sindicalistas-asesinados-Colombia_0_630537042.html

El Espectador. (2012, May 24). 'Un muerto y cuatro heridos en Atentado contra un sindicalista en Cali'. http://www.elespectador.com/noticias/nacional/articulo-348402-un-muerto-y-cuatro-heridos-atentado-contra-un-sindicalista-cali

El Heraldo. (2009, September 10). 'ONU pide investigar presencia paramilitar'. http://www.elheraldo.hn/especiales/honduras%20en%20contra%20de%20la%20ilegalidad%20del%202004%20de%20septiembre%20de%202009/ediciones/2009/10/09/noticias/onu-pide-investigar-presencia-paramilitar

El Tiempo. (2009, September 14). 'Estarian reclutando ex-paramilitares para que viajen como mercenarios a Honduras'. http://www.eltiempo.com/colombia/justicia/estarian-reclutando-ex-paramilitares-para-que-viajen-como-mercenarios-a-honduras_6086547-1

— (2010, September 17). 'Partido Liberal rechazó declaraciones de Piedad Córdoba y las calificó como Desafortunadas'. http://www.eltiempo.com/archivo/ documento/ CMS-7917941

— (2011, October 31). 'El DAS se acabó el día en que cumplió 58 años'. http://www.eltiempo.com/justicia/ARTICULO-WEB-NEW_NOTA_ INTERIOR-10674948.html

— (2012, April 29). 'La restitución no es una lucha entre ricos y pobres: gerente INCODER'. http://www.eltiempo.com/archivo/documento/CMS-11678683

— (2013, May 15). Condena a Cesar Perez por masacre de Segovia tardo 25 años. http://www.eltiempo.com/archivo/documento/CMS-12801023

Elwert, G. (1999). 'Markets of violence'. In G. Elwert, S. Feuchtwang and D. Neubert (eds.), *Dynamics of Violence: Processes of Escalation and De-escalation in Violent Group Conflicts.* Berlin: Duncker and Humblot, pp. 1–22.

— Feuchtwang, S. and Neubert, D. (1999). 'Introduction'. In *Dynamics of Violence: Processes of Escalation and De-escalation in Violent Group Conflicts.* Berlin: Duncker and Humblot.

ENS (Escuela Nacional Sindical). (2010). 'La coyuntura laboral y sindical hechos y cifras más relevantes 2007–2008'. http://www.renovacionmagisterial.org/comunidad/docs/may2008/ens.pdf

Equipo Nizkor. (2005, December 9). 'Paramilitaries murder 22 peasant farmers in Cesar, Northern Colombia'. http://www.derechos.org/nizkor/colombia/ doc/curumanien.html

Eseverri, C. (1979). *Diccionario etimologico de helenismos españoles.* Aldecoa: University of Virginia.

Fals, O. (1965). 'Violence and the breakup of tradition in Colombia'. In C. Velez (ed.), *Obstacles to Change in Latin America.* London: Oxford University Press, pp. 188–203.

FARC-EP (Fuerzas Armadas Revolucionarias de Colombia – Ejército del Pueblo). (1999). *FARC-EP Historical Outline.* Toronto: International Commission.

Feder, D. (2004, June 28). 'Increasing repression, US intervention, and popular opposition in Colombia: a conversation with Colombian academic journalist Alfredo Molano'. *Narco News Bulletin* 33. http://www.narconews.com/Issue 33/article1003.html

Fernández, A. (2003, October 15). 'Colombia's new age of terror: Uribe attacks human rights groups as supporters of terrorism'. *Narco News Bulletin* 31. http://www.narconews.com/Issue31/article881.html

Fernandez, B. (2014, January 27). 'Mexico's Vigilante Monster'. Al-Jazeera. http://www.aljazeera.com/indepth/opinion/2014/01/mexico-vigilante-monster-20141275442528978o.html

Ford, D. (2009, February 9). 'Peru mining security firm faces investigation'. http://www.reuters.com/article/2009/02/09idUSN09520637

Frayssinet, F. (2007, May 16). 'Brazil: Dorothy Stang sentence more than symbolic?' IPS. http://www.ipsnews.net/news.asp?idnews+37757

Fuentes, F. (2011, May 22). 'Venezuela: rural killers enjoy impunity'. Greenleft. http://www.greenleft.org.au/node/47646

Galeano, E. (1973). *Open Veins of Latin America: Five Centuries of the Pillage of a Continent*. London: Monthly Review Press.

Garcia, A. (1971). *La dialectica de la democracia*. Bogotá: Ediciones Cruz del Sur, Colombia.

Gareau, F. H. (2004). *State Terrorism and the United States: From Counterinsurgency to the War on Terrorism*. London: Zed Books.

Garzon, J. C. (2005). 'La complejidad paramilitar: una aproximación estrategica'. In A. Rangel (ed.), *El Poder Paramilitar*. Bogotá: Planeta, pp. 50–101.

Gaviria, J. O. et al. (2008). *Parapolitica: Verdades y Mentiras*. Bogotá: Planeta.

Gil Delgado, J. E. (2006, October 25). 'El trabajo infantil en Colombia'. *Monografías*. http://www.monografias.com/trabajos10/train/train.shtml?relacionados

Giraldo, J. (1996). *Colombia: The Genocidal Democracy*. Monroe, ME: Common Courage Press.

Global Security. (2012). 'El Salvador civil war'. http://www.globalsecurity.org military/world/war/elsalvador2.htm.

Golinger, E. (forthcoming). 'Venezuela: paramilitary invasion'. In J. Hristov and J. Sprague (eds.), *Political Violence and Irregular Armies: The Faces of Paramilitarism in the Americas*.

Gordon, T. (2008, June 11). 'Disaster in the making: Canada concludes its free trade agreement with Colombia'. *New Socialist*. http://www.newsocialist.org/index.php?id=1627

— (2010). 'Positioning itself in the Andes: critical reflections on Canada's relations with Colombia'. *Canadian Journal of Latin American and Caribbean Studies* 35(70), pp. 51–84.

GRAIN. (2010, March). 'Grabbing in Latin America'. http://www.grain.org/articles/?id=61

— (2011, April 18). 'It's time to outlaw land grabbing, not to make it "responsible"'. http://www.commondreams.org/view/2011/04/18

Grandin, G., and Salas, M. T. (2011, May 22). 'What the "Farc Files" really reveal'. http://www.guardian.co.uk/commentisfree/cifamerica/2011/may/10/farc-files-colombia-venezuela

Granovsky-Larsen, S. (forthcoming). 'Guatemala: terror for hire'. In J. Hristov and J. Sprague (eds.). *Political Violence and Irregular Armies: The Faces of Paramilitarism in the Americas.*

Grinspun, R. (2003). 'Exploring the links among global trade, industrial agriculture, and rural underdevelopment'. In L. North and J. Cameron (eds.), *Rural Progress, Rural Decay: Neoliberal Adjustment Policies and Local Initiatives.* Bloomfield: Kumarian Press, pp. 46–68).

Grossman, P. (2000). 'India's secret armies'. In B. B. Campbell and A. D. Brenner (eds.), *Death Squads in Global Perspective: Murder with Deniability.* New York: St. Martin's Press, pp. 261–86.

Guardian. (2012, January 27). 'Hungary's Roma clash with right-wing paramilitary group'. http://www.guardian.co.uk/world/video/2012/jan/27/hungary-roma-rightwing-paramilitary-video

Guerrero Barron, J. (2001). 'Is the war ending? Premises and hypotheses with which to view the conflict in Colombia'. *Latin American Perspectives* 28(12), pp. 12–30.

Guillen, F. (1963). *Raiz y futuro de la revolución.* Bogotá: Ediciones Tercer Mundo.

Gutiérrez, C. (2005, October 12). 'Land grabs by narco-traffickers and paramilitaries'. *Le monde diplomatique.* http://mondediplo.com/2005/10/12colombia

Hanson, J. K. and Sigman, R. (2011, September). 'Measuring state capacity: assessing and testing the options'. Paper prepared for the 2011 Annual Meeting of the American Political Science Association.

Harvey, D. (2003). *The New Imperialism.* Oxford: Oxford University Press.

Holmes, J. S., Gutierrez de Pineres, S. A. and Curtin, K. M. (2008). *Guns, Drugs, and Development in Colombia.* Austin: University of Texas Press.

Hristov, J. (2009a). *Blood and Capital: The Paramilitarization of Colombia.* Columbus: Ohio State University Press.

— (2009b, July/August). 'Legalizing the illegal: paramilitarism in Colombia's "post-paramilitary" era'. *NACLA Report on the Americas.* https://nacla.org/ files/ A04204014_1.pdf

Huggins, M. K. (1991). 'Introduction: a look South and North'. In *Vigilantism and the State in Modern Latin America: Essays on Extralegal Violence.* New York: Praeger Publishers, pp. 1–18.

— (2000). 'Modernity and devolution: the making of police death squads in modern Brazil'. In B. B. Campbell and A. D. Brenner (eds.), *Death Squads in Global Perspective: Murder with Deniability.* New York: St. Martin's Press, pp. 203–28.

Humanidad Vigente. (2011, March 15). 'Arauca: persecución a campesinos'. http://www.nasaacin.org/index.php?option=com_content&view=article&id=1778:arauca-persecusion-a-campesinos&catid=1:ultimas-noticias

HRW (Human Rights Watch). (1996, November 16). 'Las redes de asesinos de Colombia: La asociación militar-paramilitares y Estados Unidos'. http://www. hrw.org/spanish/informes/1996/colombia2.html

— (2003). 'World report 2003: Colombia'. http://www.hrw.org/wr2k3/americas4. html

— (2009, January 12). 'US: award to Uribe sends wrong message'. http://www.hrw. org/en/news/2009/01/12/us-award-uribe-sends-wrong-message

— (2010). 'Paramilitaries' Heirs: the Face of Violence in Colombia'. http://www. hrw.org/sites/default/files/reports/colombia0210webwcover_0.pdf

— (2012). 'World report 2011: Colombia'. http://www.hrw.org/world-report-2011/ world-report-2011-colombia

— (2013). 'Colombia: landmark ruling for land restitution'. http://www.hrw.org/ news/2013/02/20/colombia-landmark-ruling-land-restitution

Huhle, R. (2001). 'La violencia paramilitar en Colombia: historia, estructuras, politicas del estado e impacto politico'. *Revista del CESLA (2)*, pp. 63–81.

Hunter-Bowman, J. (2011, May 9). 'Colombia free trade deal could boost cocaine exports'. Common Dreams. http://www.commondreams.org/view/2011/05/09-11

ICFTU. (2005a, February 25). 'Colombia: unions appeal to Europe and ILO'. http:// www.icftu.org/displaydocument.asp?Index=991221674&Language=EN

— (2005b, October 18). 'Murders and death threats against trade unionists in the Americas on the rise, Colombia the bloodiest of them all'. http://h-net.msu.edu/ cgi-bin/logbrowse.pl?trx=vx&list=H-Labor&month=0510&week=c&msg=F3SI DLZDoerRvnDcG%2BqlYA&user=&pw=

ICTUR (International Centre for Trade Union Rights) (2010, September). Colombia Bulletin. www.ictur.org

INCODER (Instituto Colombiano Para el Desarollo Rural). (2012). 'Funciones'. http://www.incoder.gov.co/funciones_incoder/funciones_incoder.aspx

International Crisis Group. (2006, March 14). 'Colombia: towards peace and justice?' *International Crisis Group Latin America Report* 16. http://www.crisisgroup.org

Janicke, K. (2009, November 9). 'Colombia-Venezuela: the threat of imperialist war looms in the Americas'. http://www.venezuelanalysis.com/analysis/4922

Jessop, B. (1990). *State Theory: Putting the Capitalist State in its Place*. University Park: University of Pennsylvania Press.

Jones, A. (2008). 'Parainstitutional violence in Latin America'. *Latin American Politics and Society* 46(4), pp. 127–48.

Kaldor, M. (2007). *New and Old Wars: Organized Violence in a Global Era*. Cambridge: Polity Press.

Kalyvas, S. and Arjona, A. (2005). 'Paramilitarismo: Una Perspectiva Teorica'. In A. Rangel (ed.), *El Poder Paramilitar*. Bogotá: Planeta, pp. 25–40.

Koonings, K., and Krujit, D. (eds.). (1999). *Societies of Fear*. New York: Zed Books.

— (eds.). (2004). 'Introduction'. In *Armed Actors: Organized Violence and State Failure in Latin America*. London: Zed Books, pp. 1–4.

— (2007). 'Introduction: the duality of Latin American cityscapes'. In *Fractured Cities: Social Exclusion, Urban Violence and Contested Spaces in Latin America*. London: Zed Books, pp. 1–6.

Kovalik, D. (2010, April 1). 'US and Colombia cover up atrocities through mass graves'. *Huffington Post*. http://www.huffingtonpost.com/dan-kovalik/us-colombia-cover-up-atro_b_521402.html

Lair, E. (2003, June). 'Reflexiones acerca del terror en los escenarios de Guerra Interna'. *Revista de Estudios Sociales*, Facultad de Ciencias Sociales, Uniandes/ Fundacion Social, 15, pp. 96–120.

Lamounier, B. (1999). 'Brazil: inequality against democracy'. In L. Diamond et al. (eds.), *Democracy in Developing Countries: Latin America*. Boulder, CO: Lynne Rienner Publishers, pp. 119–70.

Larsen, S. (2002, October 21). 'Uribe's dictatorial rule suits oil companies'. *ZNet*. http://zmag.org/znet/viewArticle/11501

Leech, G. (2003, January 20). 'Colombia's neoliberal madness'. *Colombia Journal*. http://www.colombiajournal.org/colombia146.htm

— (2004). 'The war on terror in Colombia'. *Colombia Journal*. http://www.colombiajournal.org/War_on_Terror.pdf

— (2006, July 31). 'Uribe's new economic reforms benefit corporations, not Colombians'. *Colombia Journal*. http://www.colombiajournal.org/ Colombia241.htm

— (2008a). 'Distorted perceptions of Colombia's conflict'. *Columbia Journal*. http://colombiajournal.org/distorted-perceptions-of-colombia-conflict.htm

— (2008b, July 8). 'A more plausible scenario for Colombia hostage saga'. *Colombia Journal*. http://www.colombiajournal.org/colombia286.htm

— (2009, July/August). 'The oil palm industry: a blight on Afro-Colombia'. *NACLA Report on the Americas*. https://nacla.org/files/A04204032_1.pdf

Leeds, E. (2007). 'Rio De Janeiro'. In K. Koonings and D. Krujit (eds.), *Fractured Cities: Social Exclusion, Urban Violence and Contested Spaces in Latin America*. London: Zed Books, pp. 23–35.

Lefeber, L. (2003). 'Problems of contemporary development'. In L. North and J. Cameron (eds.), *Rural Progress, Rural Decay: Neoliberal Adjustment Policies and Local Initiatives*. Bloomfield: Kumarian Press, pp. 25–45.

Leon, J. (2011, September 15). 'Porque nombro Uribe a Jorge Noguera en el DAS?' *La Silla Vacía*. http://www.lasillavacia.com/historia/por-que-nombro-uribe-jorge-noguera-en-el-das-27693

Levy, B. H. (2004). *War, Evil, and the End of History*. Hoboken, NJ: Melville House.

Livingstone, G. (2004). *Inside Colombia: Drugs, Democracy, and War*. Piscataway, NJ: Rutgers University Press.

Lowy, M., Sader, E. and Gorman, S. (1985). 'The militarization of the state in Latin America'. *Latin American Perspectives* 12(7), pp. 7–40.

McGreevey, W. P. (1971). *An Economic History of Colombia, 1845–1930*. London: Cambridge University Press.

McLean, P. (2002). 'Colombia: failed, failing or just weak?' *The Washington Quarterly* 25(3), pp. 123–34.

Maps of World. 'Colombia: Political Map'. http://www.mapsofworld.com/colombia/colombia-political-map.html

Marion, G. (2002). 'US military bases and empire'. *Monthly Review* 53(10), pp. 1–14.

Marshall, G. (1998). 'Capital accumulation'. *A Dictionary of Sociology. Encyclopedia. com.* http://www.encyclopedia.com/doc/1O88-capitalaccumulation.html

Marx, K. (1847). 'Coming Upheaval'. *The Poverty of Philosophy*. In R. C. Tucker (ed.), *The Marx-Engels Reader*. New York: W. W. Norton and Company, pp. 218–19.

— (1867/1990). *Capital, Vol. I*. London: Penguin Books Ltd.

— (1847/1978). 'Wage Labour and Capital'. In R. C. Tucker (ed.), *The Marx-Engels Reader*. New York: W. W. Norton and Company, pp. 203–17.

— (1894/1978). 'Classes'. *Capital Vol. III*. In R. C. Tucker (ed.), *The Marx-Engels Reader*. New York: W. W. Norton and Company, pp. 439–42.

— (1941/1978). *The Grundrisse*. In R. C. Tucker (ed.), *The Marx-Engels Reader*. New York: W. W. Norton and Company, pp. 221–93.

— and Engels, F. (1846/1968). *The German Ideology*. Ed. S. Ryazanskaya. Moscow: Progress Publishers.

— and Engels, F. (1848/1987). *The Communist Manifesto*. Ed. J. Wayne. Toronto: Canadian Scholars' Press.

May, R. (2001). *Terror in the Countryside: Campesino Responses to Political Violence in Guatemala, 1954–1985*. Athens, OH: University Centre for International Studies.

Mazzei, J. (2009). *Death Squads or Self Defense Forces: New Paramilitary Groups Emerge and Challenge Democracy in Latin America*. Chapel Hill: University of North Carolina Press.

— (forthcoming). 'Paramilitary emergence and evolution in Chiapas, Mexico'. In J. Hristov and J. Sprague (eds.). *Political Violence and Irregular Armies: The Faces of Paramilitarism in the Americas*.

Mechoulan, D. (2011, November 1). 'Forced disappearances in Colombia'. COHA. http://www.coha.org/forced-disappearances-in-colombia

Medina, C. (1990). *Autodefensas, paramilitares y narcotráfico en Colombia: Origen, desarrollo y consolidación. El Caso de Puerto Boyacá*. Bogotá: Editorial Documentos Periodisticos.

— and Tellez, M. (1994). *La violencia parainstitucional paramilitar y parapolicial en Colombia*. Bogotá: Rodríguez Quito Editores.

Merriam-Webster Dictionary. 'Paramilitary'. http://www.merriam-webster.com/dictionary/paramilitary

Mesa, D. (1972). 'Treinta años de historia Colombiana (1925–1955)'. In D. Mesa et al. (eds.), *Colombia: Estructura política y agraria*. Bogotá: Ediciones Estrategia, pp. 41–2.

Miles, D. (2007, January 22). 'Pace points to Colombia's counter-drug efforts as possible model for Afghanistan'. *American Forces Press Service*. http://www.

blackanthem.com/News/International_21/Pace_Points_to_Colombia_s_ Counter-rug_Efforts_as_Possible_Model_for_Afghanistan_printer.shtml

Miliband, R. (1973). *The State in Capitalist Society: The Analysis of the Western System of Power*. London: Quartet Books.

Molano, A. (1988, June). 'Violencia y colonización'. *Revista Foro* (Bogotá) 6, 92.

— (2005). *The Dispossessed: Chronicles of the Desterrados of Colombia*. Chicago: Haymarket Books.

Moloney, A. (2005, June 27). 'Terror as anti-union strategy: the violent suppression of labour rights in Colombia'. *Multinational Monitor*. http://www.zmag.org/ content/ print_article.cfm?itemID=8174&sectionID=45

Moreno, M.F. (2006, February 20). 'Los patrocinadores de Uribe'. *Semana*. http:// www.semana.com/wf_InfoArticulo.aspx?idArt=92835

Murillo, M. and Avirama, J. R. (2003). *Colombia and the United States: War, Unrest, and Destabilization*. New York: Seven Stories Press.

Nicora, N. (2004, October 6). 'Murdered for standing up for public education'. *Network for Education and Academic Rights*. http://www.nearinternational.org/ alerts/colombia20041006.php

NODAL (Noticias de Latinoamerica y el Caribe) (2014, January 21). Marcha Patriótica denuncia asesinato de 26 miembros y se plantea disolución. http://www.nodal. am/index.php/2014/01/colombia-marcha-patriotica-denuncia-asesinato-de-26-miembros-y-se-plantea-disolucion

Nuevo Arco Iris. (2011). *La Economia de los Paramilitares*. Bogotá.

O'Connor, D. and Bohorquez, J. P. (2010). 'Neoliberal transformation in Colombia's goldfields: Development strategy or capitalist imperialism?' *Labour, Capital, and Society* 43(2), pp. 86–118.

ONIC (Organización Nacional Indígena de Colombia). (2007, May 28). 'Tres niños arrojados al río por el ESMAD'. http://colombia. indymedia.org/ news/2007/05/66050.php

Oquist, P. (1980). *Violence, Conflict, and Politics in Colombia*. New York: Academic Press.

Ortiz, S. R. (1973). *Uncertainties in Peasant Farming: A Colombian Case*. London: The Athlone Press.

Ortiz, U. (2009, March 11). 'Corrupción en el Incoder'. *El Espectador*. http://www. elespectador.com/columna126458-corrupcion-el-incoder

Palacio, G. A. (1991). 'Institutional crisis, parainstitutionality, and regime flexibility in Colombia: The place of narcotraffic and counterinsurgency'. In M. K. Huggins (ed.), *Vigilantism and the State in Modern Latin America: Essays on Extralegal Violence*. New York: Praeger Publishers, pp. 105–24.

Paley, D. (2011, December 12). 'Militarized mining in Mexico'. http://www. dominionpaper.ca/articles/4301

Pansters, W. and Castillo Berthier, H. (2007). 'Mexico City'. In K. Koonings and D. Krujit (eds.), *Fractured Cities: Social Exclusion, Urban Violence and Contested Spaces in Latin America*. London: Zed Books, pp. 36–56.

Pardo, R. (2000, July/August). 'Colombia's two front war'. *Foreign Affairs* 79(4), pp. 64–73.

Payne, L. A. (2000). *Uncivil Movements: The Armed Right-wing and Democracy in Latin America*. Baltimore: Johns Hopkins University Press.

PDA (Polo Democratico Alternativo). (2007, November 27). 'Polo Democratico Alternativo Declaration on Plan Colombia Phase II'. http://forcolombia.org/polostatement

Pearce, J. (1990). *Colombia: Inside the Labyrinth*. London: Latin America Bureau.

— (1998). 'From civil war to civil society'. *International Affairs* 74(3), pp. 587–615.

— (2010). 'Perverse state formation and securitized democracy in Latin America'. *Democratization* 17(2), pp. 286–306.

Pereira, A. and Davis, D. E. (2000). 'Introduction: new patterns of militarized violence and coercion in the Americas'. *Latin American Perspectives* 27(3), pp. 3–17.

Perelman, M. (2001). 'The secret history of primitive accumulation and classical political economy'. *The Commoner*. http://www.thecommoner.org

Perez-Rincon, M. A. (2006). 'Colombian international trade from a physical perspective: Towards an ecological "Prebisch thesis"'. *Ecological Economics* 59(4), pp. 519–29.

Petras, J. (2003). *The New Development Politics: The Age of Empire Building and New Social Movements*. Hants, UK: Ashgate Publishing Limited.

Pimiento, S. and Poland, J. L. (2012, May 21). 'United States sends combat commanders to Colombia'. http://www.justiceforcolombia.org/news/article/1231/united-states-sends-combat-commanders-to-colombia

Pitarque, P. (2010, April 18). 'Colombian court stands tall on US military base issue'. COHA. http://www.coha.org/colombian-court-stands-tall-on-us-military-base-issue

Pizarro, E. (2004). *Una democracia asediada: Balance y perspectivas del conflicto armado en Colombia*. Bogotá: Grupo Editorial Norma.

Prensa Latina. (2012, March 5). 'Las alarmantes cifras de la pobreza extrema en Colombia'. Prensa Latina. http://matrizur.org/index.php?option=com_content&view=article&id=20335:las-alarmantes-cifras-de-la-pobreza-extrema-en-colombia&catid=37:patria-grande&Itemid=56

Privacy International. (2004, September 19). 'Terrorism profile – Colombia'. http://www.privacyinternational.org/article.shtml?cmd%5B347%5D=x-347-68924

Procuraduría General de la Nación. (2006, February 12). 'Procuraduría sancionó a coronel del ejército por omisión para evitar masacre en Alto Naya'. http://www.procuraduria.gov.co/html/noticias_2006/noticias_057.htm

Ramírez, F. (2005). *The Profits of Extermination: How US Corporate Power is Destroying Colombia*. Monroe, ME: Common Courage Press.

Randall, S. J. (1992). *Colombia and the United States: Hegemony and Interdependence*. Athens: University of Georgia Press.

Rangel, A. (2005). 'Adonde van los paramilitares?' In *El Poder Paramilitar*. Bogotá: Planeta, pp. 1–24.

Rebelion. (2009, May 23). 'Viaje a los hornos crematorios que construyeron los paramilitares en Norte de Santander'. http://www.rebelion.org/noticia. php?id=85809

Reinhardt, N. (1988). *Our Daily Bread: The Peasant Question and Family Farming in the Colombian Andes.* Berkeley: University of California Press.

Rejali, D. (1991). 'The discourse about violence: Introduction'. In M. K. Huggins (ed.), *Vigilantism and the State in Modern Latin America: Essays on Extralegal Violence.* New York: Praeger Publishers, pp. 125–6.

Remmer, K. L. (2003). 'Elections and economics in contemporary Latin America'. In C. Wise and R. Roett (eds.), *Post-Stabilization Politics in Latin America.* Washington: Brookings Institution Press, pp. 31–55.

República de Colombia. (1946). 'Presidencia de la República, Un año de gobierno, 1945–1946: Discursos y otros documentos'. Bogotá: Imprenta Nacional.

Restrepo, J. D. and Franco, V. L. (2007). 'Dinamica reciente de reorganización paramilitar en Colombia'. *Controversia* 189, pp. 63–95.

Restrepo, L. A. (2004). 'Violence and fear in Colombia: fragmentation of space, contraction of time and forms of evasion'. In K. Koonings and D. Krujit (eds.), *Armed Actors: Organized Violence and State Failure in Latin America.* London: Zed Books, pp. 172–85.

Reyes Echandia, A. (1991). 'Legislation and national security in Latin America'. In M. K. Huggins (ed.), *Vigilantism and the State in Modern Latin America: Essays on Extralegal Violence* New York: Praeger Publishers, pp. 142–55.

Rich, P. (1999). 'The emergence and significance of warlordism in international politics'. In *Warlords in International Relations.* London: Macmillan Press, pp. 1–16.

Richani, N. (2002). *Systems of Violence: The Political Economy of War and Peace in Colombia.* Albany: State University of New York Press.

— (2007). 'Caudillos and the crisis of the Colombian state: fragmented sovereignty, the war system and the privatization of counterinsurgency in Colombia'. *Third World Quarterly* 28(2), pp. 403–17.

— (2010). 'Colombia: predatory state and rentier political economy'. *Labour, Capital and Society* 43(2), pp. 120–41.

Robinson, Major T. P. (2001). 'Twenty-first century warlords: diagnosis and treatment?' *Defense Studies* 1(1), pp. 121–45.

Robinson, W. (2003). *Transnational Conflicts: Central America, Social Change, and Globalization.* New York: New Left Books.

— (2004). *A Theory of Global Capitalism: Transnational Production, Transnational Capitalists, and the Transnational State.* Baltimore: Johns Hopkins University Press.

Rojas, A. (2012). *Contrainsurgencia Sin Límites en Países Estratégicos: el Paramilitarismo en Colombia* (unpublished).

Rojas, D. M. (2007). 'Plan Colombia II: Mas de lo mismo?' *Colombia Internacional* 65. http://www.scielo.org.co/scielo.php?pid=S0121-5612007000100002&script=sci_arttext

Rojas, P. (2009, November 13). 'Rural inequality: historical curse or state policy?' *Colombia Reports*. http://colombiareports.com/opinion/the-colombiamerican/6865-rural-inequality-historical-curse-or-state-policy.html

Romero, M. (2000). 'Changing identities and contested settings: regional elites and the paramilitaries in Colombia'. *International Journal of Politics, Culture, and Society* 14(1), pp. 51–69.

— (2003). *Paramilitares y autodefensa*. Bogotá: Planeta.

— (2007). *Parapolitica: La Ruta de la Expansion Paramilitar y Los Acuerdos Politicos*. Bogotá: Corporacion Nuevo Arco Iris.

Ron, J. (2000). 'Territoriality and plausible deniability: Serbian paramilitaries in the Bosnian war'. In B. B. Campbell and A. D. Brenner (eds.), *Death Squads in Global Perspective: Murder with Deniability*. New York: St. Martin's Press, pp. 287–312.

Rosenbaum, H. R. (1974). 'Vigilantism: an analysis of establishment violence'. *Comparative Politics* 6(4), pp. 541–70.

Rozema, R. (2007). 'Medellín'. In K. Koonings and D. Krujit (eds.), *Fractured Cities: Social Exclusion, Urban Violence and Contested Spaces in Latin America*. London: Zed Books, pp. 57–70.

Rudqvist, A. (1983). 'La Organización Campesina y La Izquierda: ANUC en Colombia 1970 and 1980'. *Informes de Investigación No.1*. Centro de Estudios Latinoamericanos. http://www.kus.uu.se/pdf/publications/Colombia/ Organizacion_Campesina_y_ANUC.pdf

Sader, E. (1977). *America Latina bajo la hegemonia militar*. Paris: Centro de Informacion de America Latina, Université de Paris.

Safford, F. and Palacios, M. (2002). *Colombia: Fragmented Land, Divided Society*. New York: Oxford University Press.

Salas, C. (2010, January 28). 'Paramilitaries in their DNA'. http://www.cipcol. org/?p=1297

Sánchez, G. (2001). 'Introduction: problems of violence, prospects for peace'. In C. Berquist, R. Penaranda and G. Sánchez (eds.), *Violence in Colombia 1990–2000*. Wilmington, DE: Scholarly Resources Inc., pp. 1–38.

Sánchez-Garzoli, G. (2011, June 8). 'ACSN outraged by murder of Afro-Colombian displaced leader'. WOLA. http://www.wola.org/publications/acsn_outraged_by_murder_of_afro_colombian_displaced_leader

Santa, E. (1955). *Sociologia política de Colombia*. Bogotá: Editorial Iquiema.

Scherlen, R. G. (2009, January 7–10). 'The Colombianization of Mexico? The Evolving Mexican Drug War'. Presented at the Southern Political Science Association Conference, New Orleans, LA. http://libres.uncg.edu/ir/asu/f/Scherlen_Renee_2009_the_colombianization.pdf

Semana. (2004, December 19). 'Corrupción en la DEA: Un documento secreto del Departamento de Justicia de Estados Unidos revela aterradores casos de delincuencia de agents de la DEA en Colombia'. http://semana.terra.com.co/opencms/opencms/Semana/articulo.html?id=92392

— (2005a, April 24). 'Los tentáculos de las AUC'. http://semana.com/ wf_InfoArticulo. aspx?IdArt=86215

— (2005b, April 24). 'Los "chepitos" de la costa'. http://semana.com/ wf_InfoArticulo. aspx?IdArt=86217

— (2005c, April 24). 'Los nuevos cacíques'. http://semana.com/ wf_InfoArticulo. aspx?IdArt=86218

— (2005d, February 10). 'La negociación con los paras'. http://semana.com/ wf_ InfoArticulo.aspx?IdArt=80564

— (2006, April 3). 'Habla Jorge 40: El último jefe de las autodefensas se desmoviliza y al mismo tiempo advierte que el paramilitarismo no se acabará'. http://www. semana.com/wf_InfoArticulo.aspx?IdArt=93074

— (2007a, August 25). 'Extorsiones, asesinatos y narcotrafico'. http://www.semana. com/wf_InfoArticulo.aspx?IdArt=105900

— (2007b, September 8). 'Los 40 principales'. http://www.semana.com/ wf_ infoarticulo.aspx?IdArt=106114

— (2008, November 1). 'La historia detras del remezón'. http://www.semana.com/ noticias-nacion/historia-detras-del-remezon/117295.aspx

— (2009a, March 14). 'Los estan matando'. http://www.semana.com/noticias-nacion/ estan-matando/121735.aspx

— (2009b, May 14). 'Águilas Negras envían amenazas documentadas'. http://www. semana.com/noticias-conflicto-armado/aguilas-negras-envian-amenazas-documentadas/123963.aspx

— (2009c, May 20). 'El Hedor se Toma la UIS'. http://www.semana.com/noticias-opinion-on-line/hedor-toma-uis/124212.aspx

— (2009d, June 25). 'A un año de Jaque ... el Salto Estratégico'. http://www.semana. com/noticias-seguridad/ano-jaque-salto-estrategico/124665.aspx

— (2010a, January 30). 'Politicamente incorrectos'. http://www.semana.com/noticias-nacion/politicamente-incorrectos/134341.aspx

— (2010b, February 2). 'Este caso se suma a los más de 40 uniformados que han sido absueltos por esta misma razón'. http://www.semana.com/justicia/vencimiento-terminos-libertad-otro-militar-involucrado-falsos-positivos/134476-3.aspx

— (2010c, May 19). 'La larga lista de víctimas de la vereda La Alemania en San Onofre'. http://www.semana.com/noticias-nacion/larga-lista-victimas-vereda-alemania-san-onofre/139088.aspx

— (2011a, May 28). 'Con licencia para despojar'. http://www.semana.com/ nacion/ licencia-para-despojar/157542-3.aspx

— (2011b, May 28). 'In Memoriam'. http://www.semana.com/nacion/ in-memoriam/157528-3.aspx

— (2011c, June 4). 'Neoparamilitares?' http://www.semana.com/ nacion/ neoparamilitares/157914-3.aspx

— (2011d, August 13). 'A las puertas del infierno'. http://www.semana.com/ especiales/ puertas-del-infierno/162302-3.aspx

— (2012a, January 7). 'Las Bandas, Gran Desafío'. http://www.semana.com/nacion/bandas-gran-desafio/170022-3.aspx

— (2012b, January 14). 'La Maquina Neoparamilitar'. http://www.semana.com/nacion/maquinaria-neoparamilitar/170348-3.aspx

— (2012c, February 11). 'Yo Conoci la Maldad'. http://www.semana.com/nacion/articulo/yo-conoci-maldad/253273-3

— (2013, July 2). 'Llegara el ELN a la mesa de Negociaciones?' http://www.semana.com/nacion/articulo/llegara-eln-mesa-negociaciones/349381-3

— (2014a, January 18). 'Estos son los paras de Mexico'. http://www.semana.com//mundo/articulo/michoacan-esta-fuera-de-control-por-presencia-de-autodefensas/371220-3

— (2014b, February 8). 'Seis millones de victimas deja el conflicto en Colombia'. http://www.semana.com//nacion/articulo/victimas-del-conflicto-armado-en-colombia/376494-3

Semana Multimedia. (2012). 'Violencia en cifras'. http://www.semana.com/ wf_multimedia.aspx?idmlt=93

Shivji, I. (2005, October 20). 'Primitive accumulation of wealth means reaping without sowing'. *Pambazuka News*. http://www.pambazuka.org/en/category//29962

SINALTRAINAL (Sindicato Nacional de Trabajadores de la Industria de Alimentos) (2013, November 11). 'Asesinado Sindicalista de la Nestle en Colombia'. http://www.sinaltrainal.org/index.php/empresas20/nestl%C3%A919/3560-asesinado-sindicalista-de-la-nestle-en-colombia

Sluka, J. A. (2000). 'Introduction'. In *Death Squads: The Anthropology of State Terror*. Philadelphia: University of Pennsylvania Press.

Sosa Elízaga, R. (2013). 'Facing Global Inequality: A Proposal for Sociological Debate'. International Sociological Association. http://www.isa-sociology.org/congress2014/facing-inequality.htm

Spencer, D. (2001, December). *Colombia's Paramilitaries: Criminals or Political Force?* Strategic Studies Institute.

Sprague, J. (2012). *Paramilitarism and the Assault on Democracy in Haiti*. New York: Monthly Review Press.

Stokes, D. (2005). *America's Other War: Terrorizing Colombia*. London: Zed Books.

Tate, W. (2001). 'Into the Andean quagmire: Bush II keeps up march to militarization'. *NACLA Report on the Americas* 35(3), pp. 45–50.

Taussig, M. (2003). *Law in a Lawless Land: Diary of a Limpieza in Colombia*. New York: New Press.

Territorio Chocoano (2011, August 30). 'Condenan a dos palmicultores por desplazamiento forzado en el Choco'. http://www.territoriochocoano.com/secciones/orden-publico/1830-condenan-a-dos-palmicultores-por-desplazamiento-forzado-en-el-choco.html

Tilly, C. (2003). *The Politics of Collective Violence*. Cambridge: Cambridge University Press.

— (2005). *Regimes and Repertoires*. Chicago: University of Chicago Press.

Tobon, W. R. (2005). 'Autodefensas y Poder Local'. In A. Rangel (ed.), *El Poder Paramilitar*. Bogotá: Planeta, pp. 145–80.

Toledo, R., Gutierrez, T., Flounders, S. and McInerney, A. (eds.). (2004). *War in Colombia: Made in USA*. New York: International Action Center.

Torres, C. (1970). 'La violencia y los cambios socio-culturales in las areas rurales Colombianas'. In *Cristianismo y Revolución*. Mexico: Era, pp. 227–68.

Torres-Rivas, E. (1999). 'Notes on terror, violence, fear and democracy'. In K. Koonings and D. Krujit (eds.), *Societies of Fear*. New York: Zed Books, pp. 285–300.

Torrijos, V. (2010). 'Sparks of war? Military cooperation between Colombia and the US from a strategic perspective'. Real Instituto Elcano (ARI). http://www. realinstitutoelcano.org/wps/wcm/connect/d88a2e80420f33ef9d21df9a2c3a18ac/ ARI16-2010_Torrijos_Military_Cooperation_Colombia_US.pdf? MOD=AJPER ES&CACHEID=d88a2e80420f33ef9d21df9a2c3a18ac

UN (United Nations). (2010). *Human Development Report 2009: Colombia*. http:// hdrstats.undp.org/en/countries/data_sheets/cty_ds_COL.html

— (2012). Declaration of Universal Human Rights. https://www.un.org/en/documents/ udhr

UNDP (United Nations Development Programme) (2011). *Human Development Report 2011—Sustainability and Equity: A Better Future For All*. New York: United Nations Development Programme.

US Department of State. (2009, August 18). 'US–Colombia Defence Cooperation Agreement'. http://www.state.gov/r/pa/prs/ps/2009/aug/128021.htm

USLEAP (US Labour Education in the Americas Project). (2011, June). 'Murder and impunity of Colombian trade unionists'. http://www.usleap.org/files/ Colombia%20 Fact%20Sheet_June2011.pdf

Veltmeyer, H. (1997). *Neoliberalism and Class Conflict in Latin America*. London: Macmillan Press.

Verdad Abierta. (2009a). 'La radiografía de la parapolitica legislativa'. http://www. verdadabierta.com

— (2009b, April 30). 'Dos veces despojados'. http://www.verdadabierta.com/web3/ paraeconomia/1111-dos-vecesdespojados

— (2010, August 20). 'El complejo reto de la restitución de tierras'. http://www. verdadabierta.com/paraeconomia/2657-el-complejo-reto-de-la-restitucion-de-tierras

— (2011a, February 21). 'Fiscalia abre investigación a funcionarios de INCORA e Incoder'. http://www.verdadabierta.com/conflicto-hoy/rearmados/3052-fiscalia-investiga-a-trabajadores-del-incoder-y-ex-funcionarios-por-robo-de-tierras

— (2011b, May 15). 'Asi les quitaron las tierras'. http://www.verdadabierta.com/ nunca-mas/3249-asi-les-quitaron-las-tierras

—(2011c, September 7). 'La Palma y Los Paramilitares en Choco'. http://www. verdadabierta.com/component/content/article/48-despojo-de-tierras/3526-la-palma-y-los-paramilitares-en-choco

— (2012a, May 18). 'Los nexos de Angel Maya Daza con los paramilitares'. http://www.verdadabierta.com/bandera/index.php?option=com_content&id=4014

— (2012b, May 23). 'Capturan Matias Oliveros ex-alcalde de El Banco, Magdalena'. http://www.verdadabierta.com/bandera/index.php?option=com_content&id=4015

— (2012c, June). 'Cinco años de parapolitica: ¿Que tan lejos esta el fin de la parapolitica?' http://www.verdadabierta.com/Especiales/cinco-anios-parapol/El_fin_de_la_parapolitica.pdf

Vieira, C. (2008, May 14). 'Extradition of paramilitary chiefs – A blow to truth'. *Inter Press Service News Agency.* http://www.ipsnews.net/print.asp?idnews=42356

Von Trotha, T. (1999). 'Forms of martial power: total wars, wars of pacification, and raid'. In G. Elwert, S. Feuchtwang and D. Neubert (eds.), *Dynamics of Violence: Processes of Escalation and De-escalation in Violent Group Conflicts.* Berlin: Duncker and Humblot, pp. 34–43.

Waisman, C. H. (1999). 'Argentina: capitalism and democracy'. In L. Diamond et al. (eds.), *Democracy in Developing Countries: Latin America.* Boulder, CO: Lynne Rienner Publishers.

Waldmann, P. (1999). 'Societies in civil war'. In G. Elwert, S. Feuchtwang and D. Neubert (eds.), *Dynamics of Violence: Processes of Escalation and De-escalation in Violent Group Conflicts.* Berlin: Duncker and Humblot, pp. 60–6.

Wallerstein, I. (2000). 'Globalization or the age of transition? A long-term view of the trajectory of the world system'. *International Sociology* 15(2), pp. 251–67.

Warren, K. B. (2000). 'Death squads and wider complicities: dilemmas for the anthropology of violence'. In J. A. Sluka (ed.), *Death Squads: The Anthropology of State Terror* Philadelphia: University of Pennsylvania Press, pp. 226–47.

Weaver, F. S. (2000). *Latin America in the World Economy: Mercantile Colonialism to Global Capitalism.* Boulder, CO: Westview.

Wilson, M. (2010, May 4). 'Colombia: Latin America's, if not the world's, capital of internally displaced people'. COHA. http://www.coha.org/colombia-latin-america%e2%80%99s-if-not-the-world%e2%80%99s-capital-of-internally-displaced-people

Williamson, J. and Kuczynski, P. (2003). *After the Washington Consensus – Restarting Growth and Reform in Latin America.* Washington, DC: Institute for International Economics.

WNU (Weekly News Update on the Americas). (2004, April 25). 'Colombia: Unionist's family massacred'. *WNU* 743. http://www.tulane.edu/~libweb/ RESTRICTED/WEEKLY/2004_0425.txt

WOLA (Washington Office on Latin America). (2012, January 30). '"Consolidation," land restitution, and rising tensions in Montes de María'. http://www.wola.org/commentary/consolidation_land_restitution_and_rising_tensions_in_montes_de_maria_colombia

Wolf, E. R. and Hansen, E. C. (1967). 'Caudillo politics: a structural analysis'. *Comparative Studies in Society and History* 9(2), pp. 168–79.

Wood, E. M. (1981, May/June). 'The separation of the economic and the political in capitalism'. *New Left Review* I/127. http://www.newleftreview.org/?view=1597

Wolfgang, M. and Ferracuti, F. (1967). *The Sub-culture of Violence: Toward an Integrated Theory in Criminology*. London: Tavistock.

Yepes, A. (2002). 'Adjustment produced redistribution that favours the financial sector'. Social Watch Annual Report: Colombia. http://www.socialwatch.org/ en/ informeImpreso/pdfs/colombia2002_eng.pdf

Zamora, R. (2013). 'About Santos' cruel economic policies'. FARC Peace Process. http://farc-epeace.org/index.php/what-you-should-know/item/151-about-santos'-cruel-economic-policies.html

Zarembka, P. (2002). 'Primitive accumulation in Marxism, historical or trans-historical separation from means of production?' *The Commoner*. http://www.thecommoner.org

Zuleta, E. (1973). *La tierra en Colombia*. Medellín: Editorial La Oveja Negra.

— (2005). *Colombia: Violencia, democracia y derechos humanos*. Medellín: Hombre Nuevo Editores.

Interviews

All interviewees have been assigned fictitious names to preserve their anonymity.

Interview (2009, October 16). Oscar – member of a present paramilitary group. Department of Santander.

Interview (2007, February 15). Alberto – formal military officer. Dabeiba, Department of Antioquia.

Interview (2005, August 30). Fernando – formal criminal investigator for the CTI. Department of Valle del Cauca.

Index

Compiled by Sue Carlton

Page numbers followed by n refer to the notes

www.ingramcontent.com/pod-product-compliance
Lightning Source LLC
Chambersburg PA
CBHW032132020426
42334CB00016B/1127